COMMONWEALTH ECONOMIC PAPERS: No. 16

FINANCIAL INTERMEDIATION IN SMALL ISLAND DEVELOPING ECONOMIES

A Study prepared for the Commonwealth Secretariat
by Professor Maxwell J. Fry

Commonwealth Secretariat
Marlborough House
London SW1Y 5HX

September 1981

Printed and bound by CPI Group (UK) Ltd, Croydon, CR0 4YY

May be purchased from
Commonwealth Secretariat Publications
Marlborough House
London SW1Y 5HX

ISBN 0 85092 208 9

MIX
Paper from responsible sources
FSC® C013604

PREFACE

Development of a well integrated financial system to effectively mobilise and efficiently allocate resources is essential to promote economic growth. In recent years considerable attention has been focused on the process of financial intermediation, providing a link between savings and investment activities, and its impact on economic development. However, much of the work has been in the context of larger developing countries and the working of the financial system in small economies has hardly received any attention.

The enclosed study, undertaken at the request of the Secretariat, by Professor Maxwell Fry of the University of California at Irvine, is an attempt to examine the problems of and policies for financial intermediation in small economies. The study contains an examination of the role of financial institutions in small economies, measures for their efficient working and an appropriate policy frame, along with an analysis of financial structure in eleven countries. It is hoped that the study will be useful to researchers and policy makers and will lead to further analytical work in this important area. The views expressed in this study are those of the author and do not necessarily reflect those of the Commonwealth Secretariat.

Bimal Jalan
Director, Economic Affairs

CONTENTS

FINANCIAL INTERMEDIATION IN SMALL ISLAND DEVELOPING ECONOMIES

1. Introduction

The purpose of this paper is to provide a background survey of financial intermediation in small island developing economies. An attempt to achieve this aim is made here by examining the roles of financial intermediaries, measuring the efficiency with which they perform these roles, describing briefly the financial intermediaries operating in a sample of 11 small island developing economies, discussing the part played by foreign financial institutions, analysing advantages and disadvantages of alternative financial systems for these economies, and, finally, outlining the elements of a monetary policy strategy designed to promote effective and efficient domestic resource mobilisation and allocation in small island developing economies.

At the outset, 12 countries were chosen for analysis and illustration. One of these - Antigua - was dropped for lack of data. The remaining 11 are listed with their 1979 populations and per capita gross national products (GNP) in Table 1. They were selected on the basis of size (population under 5 million), geographical spread (four from the Pacific, three from the Caribbean, two from Southeast Asia, and two from the Indian Ocean), and data availability.

On average, these small island developing economies have a population of 1.1 million and a per capita income of $2,730. However, populations range from 65,000 in the Seychelles to 4.7 million in Hong Kong. Per capita incomes vary from $200 in the Maldives to $4,000 in Hong Kong. Given this degree of variation in size and income, it is no surprise to find the economies of these countries are far from homogeneous. About the only common economic characteristic is openness. All these economies have relatively high ratios of foreign trade (imports plus exports) to GNP.

The author wishes to thank Warren Coats, Vicente Galbis, Tony Hughes, Deena Khatkhate, Ian McCarthy, Rupert Mullings and Yuzuru Ozeki for their help.

TABLE 1

1979 Population, Per Capita GNP and Aggregate GNP in Sample Countries

Country	Population (in thousands)	Per Capita GNP (in U.S. dollars)	GNP (in millions of U.S. dollars)
Bahamas	231	2,780	642
Barbados	253	2,400	607
Fiji	618	1,690	1,044
Hong Kong	4,671	4,000	18,684
Maldives	149	200	30
Papua New Guinea	3,000	650	1,950
St. Lucia	122	780	95
Seychelles	65	1,400	91
Singapore	2,368	3,820	9,046
Solomon Islands	219	440*	96
Western Samoa	158	420*	66
Total	11,854	2,729	32,351

Source: World Bank Atlas, 1980

* Estimate from alternative source.

In general, the poorest countries in this sample, i.e., the Maldives, Papua New Guinea, the Solomon Islands and Western Samoa, rely on agriculture (including, of course, fishing) for the majority of their income and employment. The middle income group, i.e., Bahamas, Barbados, Fiji, St. Lucia and Seychelles, produce a significant and growing proportion of their GNP in the tourism sector. In several of these countries, agriculture is a minor source of income and employment. Finally, the two higher income countries, Hong Kong and Singapore, have diversified economies with sizable manufacturing and service sectors.

Financial systems in this sample of countries are just as varied as their economic structures. They range from one of the most primitive monetary systems in the world, that of the Maldives, to two of the most advanced, those of Hong Kong and Singapore. In all but these latter two countries and Papua New Guinea, the financial sectors are dominated by branches or subsidiaries of foreign commercial banks. Indeed, in five of the smaller countries (as measured by the size of GNP) - Bahamas, the Maldives, Seychelles, the Solomon Islands and Western Samoa - there are no purely domestic commercial banks.

It is no coincidence to find a strong negative correlation between economy size and the preponderance of foreign banks in a country's financial sector. It reflects the basic fact that there are important economies of scale in the banking industry. These can be exploited by multinational banks even in the smallest economy, provided only that the economy is large enough to support one bank branch. The critical economy size of a bank branch's viability is much smaller than the cutoff point for a de novo bank's viability. This is simply because a branch has recourse to head office services, whereas a de novo bank must set up those services itself within the economy.

2. Role of Financial Intermediaries

Financial intermediaries perform two major economic functions in almost all economies. First, they create money and administer the payments mechanism. In most economies today, a central bank or monetary authority issues currency and depository institutions supply deposit money. Financial intermediaries administer a country's payments mechanism by providing currency notes of desired denominations when and where they are wanted and by transferring deposits, e.g., on instructions in the form of cheques.

Money's primary function is as a medium of exchange. It also serves as a unit of account, a store of value and a standard of deferred or future payment. Money emerged when and only when its use could reduce transactions costs by more than its cost of adoption. The benefits of money over barter transactions are far too well known to bear repeating. Less obvious, perhaps, is the fact that different monies perform their functions more or less efficiently.

Transactions costs are reduced most by a money whose value remains stable over time and which provides an efficient payments mechanism. Inflation erodes several of money's attributes. And money is a less efficient means of payment than it could be if, for example, there is insufficient small change, as seems to be the case frequently in Italy, if notes are so worn out that they disintegrate easily, as is the case in a number of developing countries, if counterfeit notes abound, or if deposit money cannot be transferred accurately and speedily from one party to another. Even some of the richer developing countries, e.g., Turkey, do not possess national cheque clearing facilities. Deposit transfer from one region of the country to another can be an expensive and slow procedure.

Financial intermediaries do not administer a country's payments mechanism efficiently if they have failed to develop a cheap, quick and safe method for interregional payments. For all the countries considered here, the same point applies to international payments.

The supply of money and the administration of the payments mechanism is not costless. Efficiency must, therefore, be measured in terms of the benefit/cost ratio. On this criterion, the Maldives may possess the most efficient, albeit the most primitive, system. Clearly, supplying money and administering the payments mechanism incurs social or resource costs. The resource cost of a commodity or full-bodied money equals the total value of the money supply. The costs of producing and maintaining fiat paper money rarely exceeds 5 per cent annually of the value of notes outstanding. They comprise, in the main, costs of replacing worn notes, adding additional notes, and preventing forgery. The costs solely of supplying and maintaining deposit money are far lower. They are only the bookkeeping costs. The domestic or national resource cost of using a foreign fiat money, as in the case of the Maldives' use of the U.S. dollar as money in its tourist areas, is again the total value of the foreign currency in circulation. From the national viewpoint, resource costs of commodity and foreign fiat money are identical.

The resource costs of administering the payments mechanism include the value of resources used up in the process of providing currency of desired denominations when and where it is wanted and in effecting deposit transfers. The Federal Reserve System, for example, incurs resource costs greater than the GNPs of several of the sample countries in running the national cheque clearing system in the U.S. Legal and regulatory restraints have impeded the introduction of a nationwide electronic funds transfer system, a much more efficient method than cheque clearing for transferring deposits.

For various reasons, a country may not possess the most efficient money and payments system, as measured by the highest attainable benefit/cost ratio. The government may be using money issue as a stop-gap, inefficient revenue source. There may be legal and/or regulatory constraints preventing the adoption of technological innovations, such as electronic funds transfer. A country may choose to produce its currency notes domestically, despite lower costs of notes printed abroad. Similarly, foreign banks could be excluded in favour of indigenous enterprise, despite the fact that multinational banks might bring in technical know-how at very low marginal cost, stimulate competition and facilitate the inflow of foreign capital [Grubel (1977, pp.357-358)]. Infant-industry, dependency or nationalistic arguments would be used to justify the deliberate choice of less than maximum possible economic efficiency in such a case.

Inefficiency may also be unintentional. This is likely to occur with respect to the supply and maintenance of deposit money when the deposit industry is not behaving competitively. Uncompetitive behaviour may be the result of economies of scale: in a very small economy, the banking industry may simply be a natural monopoly. More typically, however, uncompetitive behaviour is caused by reserve requirements and/or interest rate controls.

Reserve requirements and binding loan rate ceilings impose a private cost on deposit suppliers. These involve no resource cost. Hence, private total costs of supplying deposits will exceed the resource or social costs and the supply of deposit money will be suboptimal. Even if required reserve ratios are deemed necessary for prudential or monetary policy purposes, the welfare distortions can be removed completely by paying a competitive interest rate on required reserves [Fry (1979a, p.641)]. The main welfare costs of deposit and loan rate ceilings spring not from their effect on the supply of deposits but rather from their impact on financial intermediation between savers and investors. It is to this second major function of financial intermediaries that this paper now turns.

Intermediating between savers and investors differentiates financial institutions from all other business enterprises. On the one hand, financial intermediaries' assets consist predominantly of financial claims or financial instruments, i.e., claims against other economic units or ownership in them. On the other hand, financial intermediaries offer their own financial instruments to the public and to other economic enterprises. Banks offer deposits - passbook entries or deposit receipts which represent claims against the bank. Other financial intermediaries offer insurance, pensions, bonds, etc. In each case, a claim is created against the issuer. But the claim may be contingent upon special conditions - death or an accident, reaching retirement age, etc.

Financial intermediaries must attract lenders (depositors or savers) and borrowers (investors) by offering financial claims which are more attractive to savers than those offered directly by investors, and by offering more attractive loan arrangements than investors can get directly from savers. Financial intermediaries can raise the net return to savers and lower the gross cost to investors through specialisation and by reaping the economies of scale in financial transactions, information gathering and portfolio management.

Consider the situation without financial intermediaries. The market rate of interest might be 10 per cent. However, the lender (saver) must subtract from that 10 per cent perhaps 2 per cent for the costs of searching out a suitable borrower and another 2 per cent as a risk premium (the extra interest payment required to compensate for risk). The borrower is faced with costs in addition to the interest payment - the search cost of finding a lender, e.g., 3 per cent, so raising the gross cost to 13 per cent.

Financial intermediaries reduce search costs for both lenders and borrowers by specialising in an activity subject to economies of scale. They reduce risk to lenders through portfolio diversification, also subject to scale economies. Furthermore, financial intermediaries can use the law of large numbers to offer highly liquid financial claims to lenders while lending to borrowers at long term [Khatkhate and Riechel (1980, pp.510-511)]. This law is also relevant to the provision of insurance services by financial intermediaries.

Returning to the numerical example above, financial intermediaries might reduce both borrowers' and lenders' search costs to 1 per cent, and lenders' risk premium to 1 per cent. If financial intermediaries offered lenders an interest rate of 9 per cent and charged borrowers 11 per cent, lenders' risk-adjusted net yield would be increased from 6 to 7 per cent, while borrowers' gross costs would fall from 13 to 12 per cent. The 2 per cent spread between borrowing and lending rates must cover the financial intermediaries' costs and still leave something over as profit.

The effect of financial intermediation on the volume of saving and investment is illustrated in Figure 1. Without financial intermediaries, saving and

FIGURE 1

Effects of Financial Intermediation on Saving and Investment

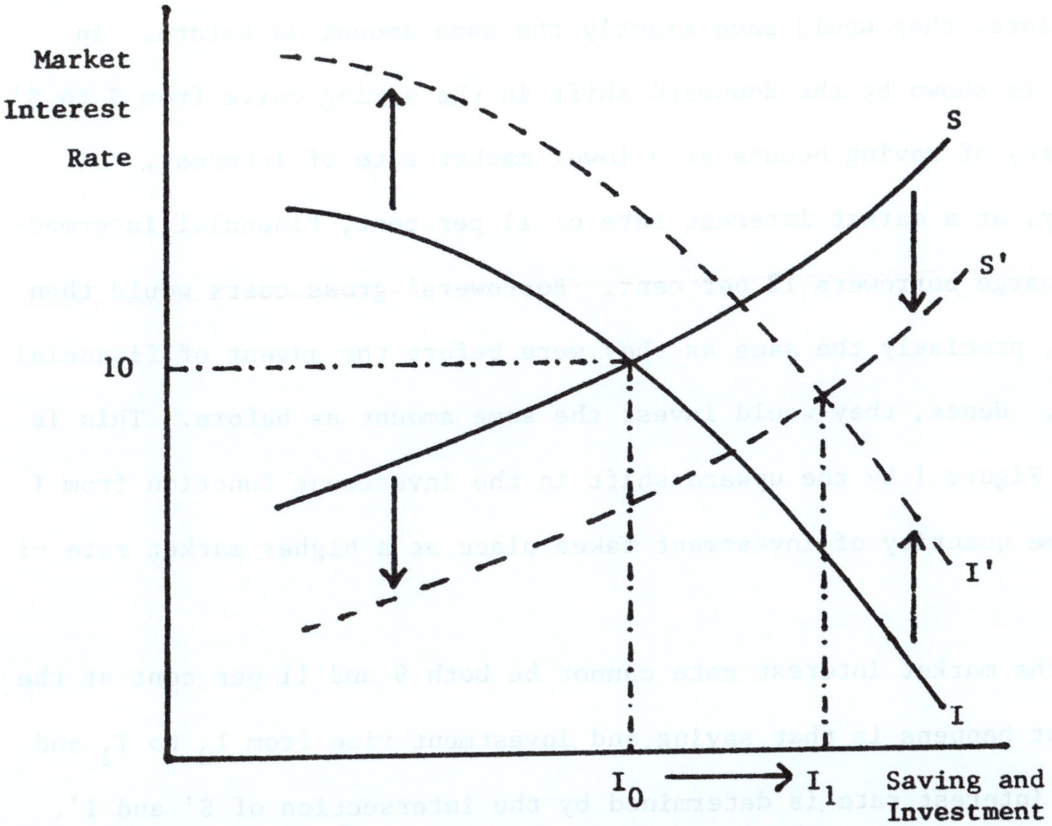

The introduction of financial intermediaries increases the market rate of interest for a given quantity of investment and decreases the market rate needed to draw forth the same quantity of saving. At a fixed market interest rate, financial intermediation increases both saving and investment.

investment of I_0 would take place at the market interest rate of 10 per cent. At a market interest rate of 9 per cent, financial intermediaries would offer savers 6 per cent. In this case, savers would be getting exactly the same risk-adjusted net return as they were before the introduction of financial intermediaries. Therefore, they would save exactly the same amount as before. In Figure 1, this is shown by the downward shift in the saving curve from S to S' - the same quantity of saving occurs at a lower market rate of interest.

Conversely, at a market interest rate of 11 per cent, financial intermediaries would charge borrowers 12 per cent. Borrowers' gross costs would then be 13 per cent, precisely the same as they were before the advent of financial intermediaries. Hence, they would invest the same amount as before. This is illustrated in Figure 1 by the upward shift in the investment function from I to I' - the same quantity of investment takes place at a higher market rate of interest.

Clearly, the market interest rate cannot be both 9 and 11 per cent at the same time. What happens is that saving and investment rise from I_0 to I_1 and the new market interest rate is determined by the intersection of S' and I'. Depending on the relative interest-elasticities of saving and investment, the new equilibrium market interest rate will be above, below or exactly at 10 per cent.

Financial intermediaries emerge and survive if and only if their indirect claims can compete successfully with the market for direct claims. This is dependent, in turn, on their ability to reduce search costs and risk, on the one hand, and the costs of so doing, on the other. As in the case of money supply, efficiency - in this case of financial intermediation - is measured by the benefit/cost ratio. It is to this topic that the paper turns next.

3. Efficiency of Resource Mobilisation by Financial Intermediaries

In practice, the benefit/cost ratio of a country's financial intermediaries cannot be measured. There is no direct indicator of total benefits. Therefore, various indirect methods of evaluating efficiency must be used. *Ceteris paribus*, unit resource costs of intermediation between savers and investors would be associated negatively with efficiency. However, it is crucial here to measure resource costs of transferring funds from original savers to final investors. Unit resource costs per financial intermediary may be very misleading if, as is the case in India, for example, there is much financial layering, i.e., one financial intermediary borrowing from another financial intermediary which, in turn, borrows from yet another. In this case, unit resource costs of each financial intermediary must be summed to produce the total intermediation cost between savers and investors.

Taking financial layering into account, costs of financial intermediation might be measured at the micro level by assessing the resource costs associated with lending for new investment from an increase in available funds, i.e., from saving. In practice, total operating costs - wages, depreciation, intermediate input costs, e.g., computer expenses, advertising, etc. - as a percentage of total earning assets may serve as a reasonable proxy. *Ceteris paribus*, the lower this percentage, the smaller will be the spread between net returns to savers and gross costs to investors. As shown in the previous section, higher net returns to savers and lower gross costs to investors may increase both saving and investment and, hence, the rate of economic growth.

Operating costs as a percentage of total earning assets can be calculated from balance sheets and income statements usually published annually by all financial intermediaries. Time series data for one particular country may indicate efficiency trends. For example, the following operating costs were calculated for all commercial banks in Turkey: 1967, 6.56 per cent; 1972, 7.43 per cent; 1977, 8.54 per cent. The comparable 1976 figure for all insured banks in the U.S. was 3.39 per cent. Operating costs of Turkish banks in 1977 were two and a half times those of American banks in 1976.

The explanation for the high and rising bank operating costs in Turkey compared to the U.S. lies in: (a) interest rate ceilings; (b) an oligopolistic and cartelised industry; (c) accelerating inflation; and (d) high and rising reserve requirements. Deposit rates of interest administratively fixed below their market equilibrium levels cause banks to substitute nonprice for price competition. In Turkey, there has been large bank expenditure on advertising and opening new bank branches. As inflation accelerated in the 1970s (from under 10 per cent in every year between 1960 and 1970 to about 100 per cent in 1980), the gap between the free competitive market equilibrium deposit rate and the fixed ceiling widened. The result was even higher levels of expenditure on nonprice competition.

Expenditure by depository institutions on nonprice competition has nowhere been valued by depositors at par with interest payments. The evidence comes from money demand estimates which invariably show that real money demand is associated positively with the real deposit rate of interest, i.e., the nominal deposit rate minus expected inflation. Money is always defined here broadly to include savings, time and post office deposits, as well as currency in circulation and sight deposits.

Nonprice competition incurs resource costs while price competition in the form of higher deposit rates does not. Interest is a transfer payment and involves no resource cost. This distinction would not matter from the welfare standpoint were the resource costs of nonprice competition valued at par by depositors. Since, clearly, they are not, there is resource misallocation when depository institutions are forced to substitute nonprice for price competition.

Deposit rate ceilings tend to encourage and condone bank cartels. Overt banking cartels for interest rate setting exist in the Bahamas, Hong Kong and Seychelles, among the sample countries. Cartels tend to raise operating resource costs, so reducing the efficiency of financial intermediation:

> Many observers have remarked that the gap between rates on loans and rates on deposits is excessive in Turkey compared to other countries. High costs would appear to absorb a large part of this gap. Why are costs high? One simple answer is that costs are high because the gap is wide. As both loan and deposit rates are fixed, perfect competition does not exist. Non-price competition in the form of massive advertising expenditure, impressive buildings, etc., takes place. Furthermore, there is no incentive to be efficient. To exhibit large profits is asking the authorities to step in and reduce the gap. This is the kind of market situation in which tacit collusion to maintain high costs and to keep profits within certain limits would flourish. [Fry (1972, p.127)]

Accelerating inflation combined with fixed or sticky nominal deposit rates of interest reduces real money demand and, hence, the real volume of resources at the disposal of the financial intermediaries. The nominal volume of deposits fails to increase in step with nominal GNP, perhaps not even in step with inflation itself, as inflation accelerates. Bank operating costs, on the other hand, do tend to rise in step with nominal GNP. Hence, as a percentage of earning assets, operating costs increase.

A dramatic inflation-induced rise in bank operating costs occurred in Brazil as inflation there rose from 13 per cent in 1952 to 41 per cent in 1966. Bank operating expenses as a percentage of bank loans outstanding increased on average from 6 per cent in 1952 to 24 per cent in 1966 [Christoffersen (1968, pp.9a and 18a)]. The explanation here is identical to that provided above for the Turkish case.

Higher reserve requirements raise bank operating costs. Ignoring bank capital and excess reserves, earning assets equal deposit liabilities when required reserves are zero. Suppose resource costs of maintaining deposits are 2 per cent of total deposits and costs of servicing the asset portfolio 3 per cent. In this case, the overall bank operating cost ratio is simply 5 per cent. Now consider what happens when a required reserve ratio of 50 per cent is imposed. For the same deposit volume, the earning assets are halved. The calculated overall bank operating cost ratio is now 3 per cent for earning assets, but resource costs for maintaining a volume of deposits twice the size of the earning asset portfolio are 4 per cent of earning assets. Hence, the overall bank operating cost ratio is now 7 per cent.

In addition to operating resource costs which drive a wedge between financial intermediaries' lending and borrowing rates, there are two additional factors - reserve requirements and taxes - to be considered. In fact, required reserves are themselves a form of taxation on financial intermediation. Suppose earning assets yield an average return of 12 per cent. Again ignoring excess reserves and capital, financial intermediaries could just offer 7 per cent on deposit liabilities and break even, given resource costs of 2 per cent for deposits and 3 per cent for earning assets, as in the example above, when the required reserve ratio is zero.

Now introduce a 50 per cent reserve requirement. The resource costs rise to 7 per cent of earning assets and the net yield on earning assets of 5 per cent has to be spread over deposits twice the size of these assets. Hence, the average deposit rate is reduced from 7 to 2½ per cent. The spread between lending and borrowing rates has widened from 5 per cent (the operating cost ratio) with zero required reserves to 9½ per cent. The 50 per cent reserve requirement effectively imposes a tax on deposits of 2½ per cent.

The reserve requirement tax increases as inflation accelerates. Suppose inflation rises from zero to 10 per cent and earning assets now yield an average return of 22 per cent (12 per cent plus the 10 per cent inflation adjustment). *Ceteris paribus*, financial intermediaries will be able to offer a maximum deposit rate of 7½ per cent - 22 minus 7, the result divided by 2. The 50 per cent reserve requirement now imposes an effective tax on deposits of 7½ per cent. And it has reduced the *real* return on deposits from +2½ per cent to -2½ per cent. Naturally, savers are deterred by the substantial decline in their real return. Real money demand and, hence, the real volume of financial intermediation will fall. This, in turn, raises the operating resource cost ratio, since financial intermediation is subject to economies of scale.

Conventional taxes - interest withholding taxes, stamp duties, transactions taxes, value added taxes, profit taxes, licence fees, etc. - all widen the competitive spread between financial intermediaries' borrowing and lending rates. They, therefore, have exactly the same effect on the real volume of financial intermediation, saving and investment as do higher operating resource costs.

If one of the objectives of an economic development plan is to encourage domestic resource mobilisation, then discriminatory taxes - conventional as well as reserve requirements - on financial intermediation should be removed. At the same time, bank cartels must be destroyed and financial intermediaries made to behave competitively. The optimal competitive solution might have to be forced upon some of the sample countries' cartelised and/or oligopolistic financial systems by fixing minimum deposit rates of interest and obliging depository institutions to satisfy all deposit demand at these rates. This would be the only control needed to produce the competitive result, provided loan demand were elastic at rates above the competitive loan rates of interest [Fry (1980b, p.543)]. Minimum deposit rates are clearly preferred to licence fees as a means of tapping monopoly profit. Barbados does actually set minimum deposit rates.

The easiest way to establish and then maintain a real deposit rate approximating the competitive rate is deposit indexation. Depository institutions would be directed to adjust nominal deposit values by the change in an appropriate index at regular intervals. For most of the sample countries, the best index might be calculated from trade-weighted average government bond yields in industrialised countries adjusted for any change in the value of the domestic currency vis-a-vis this currency basket. For example, the exchange rate might start at Rs 5 to B1, where Rs represents the domestic currency and B is the trade-weighted currency basket. Over the subsequent month, foreign government bond yields average 0.75 per cent (as a monthly rate). In this case, the index moves from 100.00 to 100.75. If the exchange rate had depreciated from Rs 5 to B1 to Rs 5.10 to B1, then the index becomes 102.765, i.e., 5.10 divided by 5, the result multiplied by 100.75.

Now financial intermediaries would adjust all deposit values by the change in the index every month. A balance of Rs 10,000 is adjusted automatically to

Rs 10,276.50. Any nominal interest offered, say, on longer-term deposits would be paid on the adjusted deposit value. Depositors know in advance that they will be compensated for subsequent price changes. Hence, their own expectations regarding the future inflation rate are now irrelevant to their deposit holding decisions. And the real deposit rate equals whatever nominal rate is paid on the adjusted deposit value plus the trade-weighted average real government bond yield abroad. The latter has tended to remain remarkably stable in most industrial countries for most of this century.

Efficiency of financial intermediation is affected positively by economy size for two reasons. First, a large economy can support a large financial intermediary and there are economies of scale to financial intermediation [Baltensperger (1972)]. Second, there is more potential for competition in a larger than in a smaller economy. Whether or not that potential is realised depends on the legal and regulatory environment in which the financial intermediaries operate. All one can conclude here is that it is probably impossible under any circumstances for the Maldives to attain as great efficiency of financial intermediation as, say, Hong Kong or Singapore.

Apart from constraints imposed by economy size, however, monetary authorities can have considerable influence over financial intermediaries' operating efficiency. No matter how small the economy, competitive conditions can be simulated through minimum deposit rates. All other interest rate regulations can be dismantled.

Domestic inflation in all the sample countries is determined by exchange rate policy. In today's inflationary world, gradual appreciation of the domestic currency vis-a-vis a trade-weighted currency basket is needed to ensure price stability. Papua New Guinea, Singapore, the Solomon Islands and Western Samoa have all opted for below-average inflation rates over the past quinquennium by means of domestic currency appreciation. Aside from other benefits of price stability, there is, *ceteris paribus*, greater efficiency of financial intermediation.

Reserve requirements do not constitute an instrument of monetary control in any of the sample countries. The nominal money supply is an endogenous variable determined *inter alia* by exchange rate policy. However, reserve requirements can and do extract seigniorage from financial intermediaries subject to them. They are a discriminatory tax which increases the spread between financial intermediaries' lending and borrowing rates. In turn, this greater spread reduces the aggregate size of the financial sector and so may raise operating resource cost ratios. In addition, of course, reserve requirements ensure a divergence between private and social or resource costs and, hence, produce a suboptimal level of financial intermediation. If required reserves are deemed essential on prudential grounds, they should receive a competitive market return [Fry (1979a, p.641)].

The Maldives and the Solomon Islands have no regulations pertaining to financial intermediaries' asset portfolios. Hong Kong and Papua New Guinea set no required reserve ratios but do impose minimum liquid asset ratios. In effect, this is almost the same as paying a competitive interest rate on required reserves. Liquid assets are defined broadly in Hong Kong and the regulation has several loopholes which render it ineffective, while financial intermediaries in Papua New Guinea typically hold liquid assets far in excess of the requirements.

Of the remaining countries in the sample, all of which do impose cash reserve requirements on their financial intermediaries, only Fiji pays any interest - 3½ per cent - on required but not excess reserves held on deposit with the Central Monetary Authority. However, the reserve requirement in Western Samoa imposes no net tax on the financial intermediaries as a group because the reserves deposited with the Monetary Board are redeposited back with one of the two commercial banks. Hence, its sole impact is to redistribute income from the newer bank - the Pacific Commercial Bank - to the commercial bank which actually issues Western Samoa's currency - the Bank of Western Samoa, in which the government has 50 per cent ownership.

The remaining five countries in the sample could raise the efficiency as well as the quantity of financial intermediation from suboptimal towards optimal levels by paying a market rate of interest on required reserves, substituting a liberally-defined liquid asset ratio for the required reserve ratio, or simply abolishing reserve requirements altogether.

Only intramarginal taxes tapping purely consumer or producer surpluses, e.g., a poll tax, produce no economic distortions. And it turns out that several of the sample countries effectively impose intramarginal taxes on money holders. Table 2 gives the current structure of deposit rates of interest and the average inflation rate as measured by the consumer price index over the period 1978-1980 for eight of the sample countries. In only one of these countries - Singapore - were deposit rates determined competitively in a free market. And in only one of these countries - Singapore - were real deposit rates of interest positive on average over the period 1978-1980. All real deposit rates were negative in the seven other countries on average over this period.

TABLE 2

Deposit Rates of Interest and Inflation

(Per Cent)

Country (Date Introduced)	Deposit Maturity: Savings	3 mth	6 mth	12 mth	2-3 yr	Inflation (1978-80)
Bahamas (1980 IV)	6.39	7.74	7.98	8.11	8.56	9.1
Barbados (3/1980)	3.0-5.0	3.5-4.5	3.8-5.0	4.0-5.5	-	13.5
Fiji (9/1980)	4.5	5.5	6.25	7.0	8.0-8.5	9.2
Papua New Guinea (12/1979)	3.75	4.75-7.2	-	5.5-7.5	-	7.6
Seychelles (5/1980)	6.5-8.0	8.0	8.5	9.0	10.0	12.6
Singapore (3/1980)	7.52	8.79	8.80	8.56	-	5.8
Solomon Islands (9/1980)	4.0	5.75-6.0	6.5	-	-	9.1
Western Samoa (6/1979)	4.0	6.5	7.0	8.0	8.5	13.8

Source: Central bank publications and International Financial Statistics

Of significance here is the fact that the term structures of deposit rates in Fiji, Papua New Guinea, Seychelles and Western Samoa are steeply rising. This indicates neither strong liquidity preference nor high marginal rates of time preference - these structures are imposed solely from the supply side. What they do approximate, however, is a monopolist's profit-maximising strategy through product differentiation [Fry (1981c)].

The problem can be viewed as one of minimising the cost of generating a given real money demand [Fry (1978c)]. Given the inflation rate and the real money demand target, interest costs will be minimised by paying zero or low nominal rates on sight and short-term time deposits, for which there are no close substitutes, and higher rates on longer-term deposits, for which closer substitutes in the form of tangible inflation hedges do exist. To stabilise the aggregate real money demand in the face of volatile inflation, the nominal rates of longer-term time deposits must be adjusted continuously in step with changes in inflationary expectations. In other words, the real deposit rate on long-term deposits has to be held constant. Therefore, the term structure has to be tilted more, the higher is the expected rate of inflation.

Whether the appropriate real rate on long-term deposits is negative, positive or zero depends on the volume of real money demand to be generated. In a growing or potentially growing economy, the basic objective of accelerating economic growth would necessitate positive real long-term time deposit rates in order to maximise the real supply of domestic credit, one of the main assets backing deposit liabilities of the financial intermediaries. Positive real long-term time deposit rates also deter socially inefficient currency substitution and excessive foreign borrowing, both of which may well involve unnecessary resource costs borne entirely by the small island developing economy [Fry (1980d, pp.888-891)].

From another viewpoint, a greater inflation tax is levied on current and short-term deposits, for which there are no close substitutes, while a smaller tax is extracted from deposits for which closer substitutes do exist. Monopolistic product differentiation is designed to tap consumer surplus, in this case from savers. The tax is used in some of the sample countries as one source of credit subsidy for priority sectors. It is also absorbed in high administrative costs of the uncompetitive financial sectors found in all these countries, with the exception of Singapore.

Governments of most of the sample countries appear to behave as if they were subject to two conflicting objectives. On the one hand, they are anxious to mobilise domestic resources by offering attractive returns to savers. On the other hand, they wish to finance their own sometimes considerable borrowing requirements as cheaply as possible. Furthermore, some of these governments - excluding those of Bahamas, Hong Kong, Maldives and Singapore - clearly believe that priority groups, sectors and regions need cheap credit.

These objectives may indeed conflict when the subsidies for government and other priority borrowers are paid by deposit holders. A proportional tax on all deposits to be spread proportionally among borrowers would have no effect at all, provided that interest rates were adjusted upwards by exactly the amount of the tax/subsidy; free market interest rates would make this adjustment automatically. Were rates not adjusted, the quantity of deposits demanded, i.e., lending, would decline.

The two objectives need not conflict, however, were the tax on deposit holders levied as a poll tax or in the form of any alternative intramarginal tax which tapped only consumer surplus. The monopolistic price discrimination in terms of the present structures of deposit yields in Fiji, Papua New Guinea, Seychelles and Western Samoa do seem, in principle, capable of tapping consumer surplus without reducing aggregate real money demand. Unfortunately, the rigidity of the structure of nominal institutional interest rates produces money demand fluctuations *pari passu* with changes in inflationary expectations. Indexation of long-term time deposits to a price index would introduce the necessary flexibility in effective nominal rates to allow consumer surplus to be tapped without reducing real money demand in the face of rising inflationary expectations.

Concern over the efficiency of financial intermediation would barely be warranted were saving completely interest-inelastic. If, however, saving is responsive to the net real yield on indirect claims, then reducing the spread between financial intermediaries' lending and borrowing rates will increase saving, investment and the rate of economic growth.

Olson and Bailey (1981, p.1) show that

> [t]he case for positive time preference is absolutely compelling, unless there is an infinite time horizon with the expectation of unending technological advance combined with what we call "drastically diminishing marginal utility." This finding holds both in the positive and normative senses. *A corollary is that savings are interest elastic*. [Italics added]

The present author has conducted empirical tests on a number of developing countries. In all cases, the results are consistent with the hypothesis that national saving rates are affected positively and significantly by the real deposit rate of interest [Abe *et al*. (1977); Fry (1976a, pp.61-62; 1978d; 1979b; 1980a; 1981a; 1981b; 1982); Fry and Farhi (1979, p.197); Fry and Mason (1981)].

4. Efficiency of Resource Allocation by Financial Intermediaries

Financial intermediaries can be evaluated not only in terms of their efficiency in mobilising resources but also on the basis of their efficiency in allocating resources. The majority of governments in the sample countries evidently believe that the financial intermediaries cannot and/or do not allocate resources efficiently. In the majority of the sample countries - the exceptions include Bahamas, Hong Kong, Maldives and Singapore - governments intervene by means of selective credit policies aimed at influencing deliberately the allocation of resources by the financial intermediaries.

Selective credit policies are designed to channel credit to priority sectors, groups and/or regions at subsidised rates of interest. The objectives are to stimulate investment in priority activities and, in many cases, to redistribute income and wealth. Selective credit policies can be implemented in five different ways.

Perhaps the most typical selective credit technique is differential rediscount rates. Financial intermediaries are compensated - partially, fully or even over-compensated - for lending at subsidised rates of interest to priority borrowers by rediscounting priority loans at the central bank or monetary authority on concessional terms, as, for example, in Barbados and Seychelles. Hence, all priority credit may actually be provided by the central bank or monetary authority. This method can and often does, e.g., in Korea and Turkey, jeopardise control over domestic credit expansion.

An extensive selective credit policy implemented through the rediscount mechanism is likely to be accompanied by high reserve ratio requirements designed to reduce the commercial banks own funds available for discretionary, nonpriority lending. Here, the central bank or monetary authority's assets will constitute a relatively large proportion of the total assets of the financial sector as a whole.

The second method is direct budgetary subsidy. Negative differentials between priority loan and deposit rates of interest may be financed by explicit budget appropriations, as is the case in Fiji and the Solomon Islands.

Credit floors constitute a third device used to implement selective credit policies. The monetary authorities set minimum proportions of total credit or total deposits which must be lent by the banks to specific priority borrowers. For example, commercial banks in Korea are obliged to extend a minimum of 30 per cent of their total loans to small and medium size industry. They have also been requested to hold a minimum percentage of their time deposits in the form of National Investment Fund (NIF) bonds. The NIF then lends to priority sectors at subsidised interest rates [Bank of Korea (1978, p.15)]. In Malaysia, commercial banks are required to lend a minimum of 20 per cent of any increase in aggregate loans to the bumiputra community, 10 per cent for agricultural production, 25 per cent to manufacturing industry, and 10 per cent for individual housing loans [Bank Negara Malaysia (1979, p.133)]. The Nepalese commercial banks must use at least 7 per cent of their total deposit liabilities for lending to small scale industries. And the commercial banks in Thailand have been directed to allocate a minimum of 13 per cent of total credit to agriculture. This requirement can be satisfied by depositing the funds with the Bank for Agricultural Cooperatives. To date, this method has not been employed formally in any of the sample countries. Informal guidelines, however, have been promulgated in Papua New Guinea.

The fourth way of implementing selective credit policies is to set credit ceilings either on nonpriority lending or on the aggregate volume of loans, a technique also employed rather informally in Papua New Guinea. Overall credit ceilings are usually set in conjunction with exemptions for priority loans and/or credit floors for priority sectors. In India and Indonesia, for example, the incentive for a commercial bank to extend subsidised credit springs, in the main, from the ceilings imposed on normal lending. A subsidised loan may be a more profitable asset than excess cash reserves. However, delinquency and default frequently plague priority lending operations [Brillembourg (1981)]. Priority lending, therefore, may yield negative nominal returns.

Finally, selective credit policies can be pursued through heavy reliance on specialised financial institutions, as in Barbados, Fiji, Papua New Guinea, St. Lucia, Seychelles, the Solomon Islands and Western Samoa. Funds are extracted from nonspecialised depository institutions through reserve requirements, etc., to be channelled to priority sectors on concessional terms by government-owned specialised financial institutions. Financial layering and market segmentation have been the main effects.

Selective credit policies use interest rate ceilings and subsidies in an attempt to direct investible funds, through a nonprice rationing system, into investments which the authorities believe might not be willingly undertaken at higher interest rates. For a selective credit policy to work at all, financial markets must be kept fragmented and segmented. Otherwise, financial channels would develop expressly for re-routing subsidised credit. Aggressive pursuit of selective credit policies or credit planning tends to produce an extensive degree of financial layering. Instead of raising funds from a common pool of loanable funds, each specialised financial intermediary established to service a particular priority sector has its own special sources of funds earmarked for its own special uses [McKinnon (1980, pp.106-110)].

In India's agricultural sector, for example, there are some 123,000 active primary agricultural cooperative societies (PACSs) covering 40 million members. These PACSs borrow funds from central cooperative banks (CCBs), of which they are members. The CCBs obtain funds from deposits, investment earnings, and refinancing facilities of the state cooperative banks (SCBs). In turn, SCBs raise funds from deposits and refinancing facilities of the Agricultural Refinance and Development Corporation (ARDC). The ARDC is financed by the Reserve Bank of India, the World Bank, and the Indian government.

It turns out that all this layering decreases rather than increases the total real volume of funds available. This is because about 7 per cent of total available resources in India are absorbed in administrative costs [Datey (1978, p.ii)], despite the fact that each individual financial institution displays relatively low operating cost ratios [Bhatt (1978)]. The direct costs of institutional agricultural credit in India average about 19 per cent. Farmers pay about 12 per cent. Therefore, all administrative costs here are covered by direct government subsidies. Furthermore, the loans supplied by the government of India to the ARDC are provided on concessional terms in the first instance. Unfortunately, India provides an excellent example of inefficient financial intermediation caused, in the main, by government intervention.

The main purposes of financial layering are: (a) to direct credit to priority activities; (b) to reimburse the final lender at least partially for the subsidy. This supply-leading approach has two major drawbacks in addition to its enormous resource costs. The first problem is that the mechanism works efficiently only to the extent that all bottom/base and middle-tier institutions possess identical lending capabilities. The refinancing mechanism comes into play only after a loan has been extended. First, however, expertise is required at the final stage or at the base of the pyramid to assist the borrower in formulating the project and preparing the loan application. The base institution also needs expertise in loan evaluation techniques. Clearly, such expertise will vary greatly from one institution to another and from one region of a country to another. Hence, credit is unlikely to be spread efficiently and equitably. Its allocation is dependent to a large extent on loan officers of differing abilities in the base institutions.

The other main problem is that there is no evidence that this expensive institutional structure has produced any benefits in terms of increased agricultural productivity. Datey (1978, p.25) poses the question as to whether agricultural productivity in India has risen despite rather than because of the system of agricultural credit. For most developing countries, Datey (1978, p.30) concludes:

> The presumed cost-benefit ratio for society as a whole must be negative in most cases, for it is rare to find a situation in which subsidies for agricultural credit have resulted in any significant increase in productivity.

One alternative to financial layering would be the adoption of more innovative financial instruments with which the base and/or middle-tier financial institutions could raise funds directly from financial markets. Indeed, it would appear to be the only way of increasing the aggregate real supply of loanable funds. This alternative, however, cannot be reconciled with the system of

real supply of credit and bias factor prices in favour of capital and against labour. As a result, the smaller volume of investment which can be financed may be less efficient.

The third inconsistency lies in the inversion of deposit and loan rates of interest. Under competitive conditions, financial intermediaries would never offer deposits with higher interest rates than their lowest loan rates of similar maturity. The fungibility of financial capital ensures that some borrowing can take place at priority loan rates for the express purpose of building up deposits yielding a higher return. Clearly, this thwarts completely the objective of the selective credit policy. It merely raises the resource costs of financial intermediation between savers and investors.

The fourth inconsistency of selective credit policies is that, if successful, they increase unemployment by distorting factor prices. The present (1981) negative loan rates of interest for priority borrowers found in all the sample countries, except Bahamas, Hong Kong, the Maldives and Singapore, give an extraordinary price signal to these entrepreneurs. The message conveyed is that this priority credit is not just a free good but actually has negative value like rubbish. The result is encouragement of highly capital-intensive production techniques for any given product, of products and processes which are necessarily capital-intensive, of investments with zero economic return, and of anti-social, i.e., unproductive, hoarding.

The fifth inconsistency is that the objectives of selective credit policies must discourage saving and so reduce the aggregate real supply of investible funds. Selective credit policies invariably keep both deposit and loan rates of interest below their market equilibrium levels. Hence, the aggregate real supply of investible funds is held below its equilibrium level.

If financial institutions are to remain solvent, a concomitant of high reserve requirements, binding quantitative ceilings on normal, higher interest rate loans, and/or binding loan rate ceilings (often differentiated for different categories of borrowers) is a lower average deposit rate of interest. The competitive solution would be lower deposit rates over the entire maturity spectrum. Consequently, aggregate real money demand would be reduced. A monopolised or cartelised banking system could, as already pointed out, lower the average deposit rate by increasing the tilt in the term structure of deposit rates. However, once the cost minimising monopoly solution is in place, an increase in either the reserve requirement or the proportion of unrediscounted low interest loans must be accompanied by a general reduction in deposit rates to maintain bank solvency. The resulting decline in aggregate real money demand will be matched by a fall in real domestic credit and/or net foreign assets.

Finally, selective credit policies provide precisely the wrong signals to private sector institutional lenders. Their incentive is to lend first at the normal rate, last at a subsidised rate. Even if compensation is provided through the rediscount mechanism, administrative costs and delays may well make such recourse unattractive, as, for example, seems to be the case in Seychelles.

Perhaps the most telling indictment of selective credit policies is their tendency to reduce the supply of credit to sectors of the economy believed to be of highest priority and in most need of financial assistance. Cheap but unavailable credit is no consolation. Furthermore, the cost of credit is rarely the main constraint to productive small-scale investments. In sum, it is far from "obvious" that "... selective credit policies should constitute an integral part of the overall economic strategy for development" [Khatkhate and Villanueva (1978, p.980)]. Rather, they seem to be an ideal recipe for reducing both the quantity and quality of productive investment. In other words, they appear to reduce rather than increase the efficiency of resource allocation by financial intermediaries.

Selective credit policies tend to be based on two premises: (a) planners know best what investments should be undertaken; (b) credit allocation can ensure that those and only those investments are undertaken. The performance of several developing economies relying heavily on selective credit policies, e.g., India, over the past two decades throws strong doubt on the first assumption. The second is belied by the fact that financial capital is fungible. Overt relending is a well-documented phenomenon. Of greater significance, however, is the fact that the fungibility of financial capital can enable a farmer to take the subsidised credit for a pumpset, which he would have bought in any case, and to use his own resources thereby released to instal air-conditioning in his home. Since the air-conditioning would not have been purchased without the subsidised credit being available, it is difficult not to conclude that this loan has financed air-conditioning rather than the pumpset. It is somewhat ironic that, to the extent that this kind of fungibility is both possible and actually effected, the inefficiencies of selective credit policies outlined above are concomitantly mitigated.

The fungibility of financial capital is well illustrated by the long history of housing finance in the U.S. For many years, it has been government policy to encourage the production of housing by increasing the availability of mortgage credit. Availability is increased by government purchase of mortgages through federal agencies. But availability is reduced back again through the bond financing of the mortgage purchases. By itself, this activity of draining funds from one part of the pond to pump back into another part has been futile [Jaffee and Rosen (1978, p.933)].

On the other hand, government action and initiative in spreading amortisation, lengthening terms, mortgage insurance and subsidies have had substantial impact on the mortgage market. "The most notable changes are growth in the number and size of specialized thrift institutions that buy mortgages and in the proportion of mortgages to total liabilities of financial institutions" [Meltzer (1974, p.764)]. However, this has had no affect on housing. The ratio of housing to total assets of non-farm households in the U.S. has remained virtually constant at 25 per cent throughout this century.

What has happened, however, is that the ratio of mortgage debt to housing has risen from about 10 to 40 per cent. The ratio of mortgage debt to total liabilities rose substantially too. Cheaper mortgage debt was substituted for more expensive (less subsidised) forms of borrowing. The conclusion which can be drawn here is that specific liabilities do not finance specific assets. Asset composition and "asset purchases are independent of the form in which credit is made available" [Meltzer (1974, p.769)].

Governments of the sample countries studied here may make more and cheaper funds available to rural areas. It is questionable, however, whether this alone would increase the stock of physical capital there. Much of the increase in mortgage debt in the U.S., for example, has actually been used to finance the purchase of securities.

The composition of spending is affected by relative prices of the goods and services being bought. Subsidised mortgages in no way affect the relative price of housing. They do, on the other hand, affect the relative costs of borrowing to buy a house vis-a-vis borrowing to buy something else or vis-a-vis a cash purchase of the house. They will, therefore, influence the way in which houses (and everything else) are financed. Subsidised mortgages will encourage people to finance a larger proportion of the purchase of a house and a smaller proportion of the purchase of other things on credit. Overall, subsidised mortgages will not affect the stock of houses or even the volume of construction, except in the short run. This is because: (a) the subsidy for one person must be a tax on someone else; (b) money channelled through one particular financial intermediary must come from another intermediary or market; (c) financial resources are fungible.

If the real concern is the distribution of income and wealth, credit subsidies are an inefficient remedy. Direct subsidy of people, not goods or finance, is the efficient solution.

Selective credit policies may not be the sole rationale behind the low interest rate policies pursued in several of the sample countries. Even the government of Singapore wishes, for budgetary reasons, to keep the cost of its borrowing low. It should be recognised, however, that the social cost of a government deficit is in no way related to the cost of servicing the national debt. The latter is simply a transfer payment. Cheap finance means that the holders - direct and indirect - of government bonds receive a lower return than they otherwise would. They will, therefore, have less incentive to hold willingly such debt. In order to make government debt relatively more attractive without increasing its costs, measures have been introduced in a number of developing countries to make private debt less attractive.

The cost of all public sector expenditure is the opportunity cost of the scarce resources consumed. Clearly, this opportunity cost remains the same whether the expenditure is financed from tax revenue or borrowing. And if financed through borrowing, it cannot be reduced by lowering the interest rate on the loan. Again, the interest cost is a transfer payment. It does not use up scarce resources. Hence, the apparent advantage of cheap public sector borrowing is, for the most part, illusory.

A low interest rate policy can often be interpreted as part of a broader policy of financial restriction, a policy which encourages financial intermediaries and financial instruments from which the government can expropriate a large seigniorage and discourages all others. For example, money and the banking system are favoured and protected - reserve requirements and obligatory holdings of government bonds can be imposed to tap this source of saving at zero or low interest cost to the public sector. Private bond and equity markets are suppressed through transactions taxes, stamp duties, special tax rates on income from capital, an unconducive legal framework, etc., because seigniorage cannot be taken so easily from private bonds and equities. Interest rate ceilings are imposed to stifle competition. Foreign exchange controls, interest rate ceilings, high reserve and/or liquidity requirements, suppression or non-development of private capital markets, etc., can all increase the flow of domestic resources to the public sector without a concomitant rise in inflation or interest rates [Fry (1973)].

Successful financial restriction would extract a substantial proportion of domestic credit at low or zero rates of interest for the public sector. The initiation of effective financial restriction would have three effects on the demand for money illustrated in Figure 2 - a rightward shift in the curve, higher income and lower cost-elasticities. Income velocity of circulation would fall and then continue to decline. Taking a higher proportion of domestic credit, i.e., extracting more seigniorage, and manipulating money demand in these ways permit a greater public sector deficit to be financed at a given rate of inflation and a given, i.e., low, level of nominal interest rates. A selective credit policy fits well into the mechanism of financial restriction.

The following describes a typical case of financial restriction:

> To finance its deficit, the government [of Portugal] has largely pre-empted the supply of domestic savings by preserving a 'sheltered' market for its own bond issues. Recourse of the private sector to the domestic bond market was, moreover, effectively curtailed by maintaining the maximum interest rate for bond issues at 5 per cent.
> [Lundberg (1964, p.40)]

The ceiling on after-tax returns from private bonds in Portugal was lower than the rate offered on government bonds. Even with these interest rate ceilings on competitive financial instruments, returns on government securities were so low that virtually no voluntary purchases took place:

> In actual fact, the vast majority of the public debt bonds were taken up by the welfare institutions, the commercial banks, the Caixa Geral de Depositos and the insurance companies.
> [Banco de Portugal (1963, p.52)]

However, the seigniorage base in the form of the money supply was large and growing. Velocity of circulation in Portugal fell smoothly from 1.46 in 1962 to 1.09 in 1973.

FIGURE 2

Demand for Money under Financial Restriction

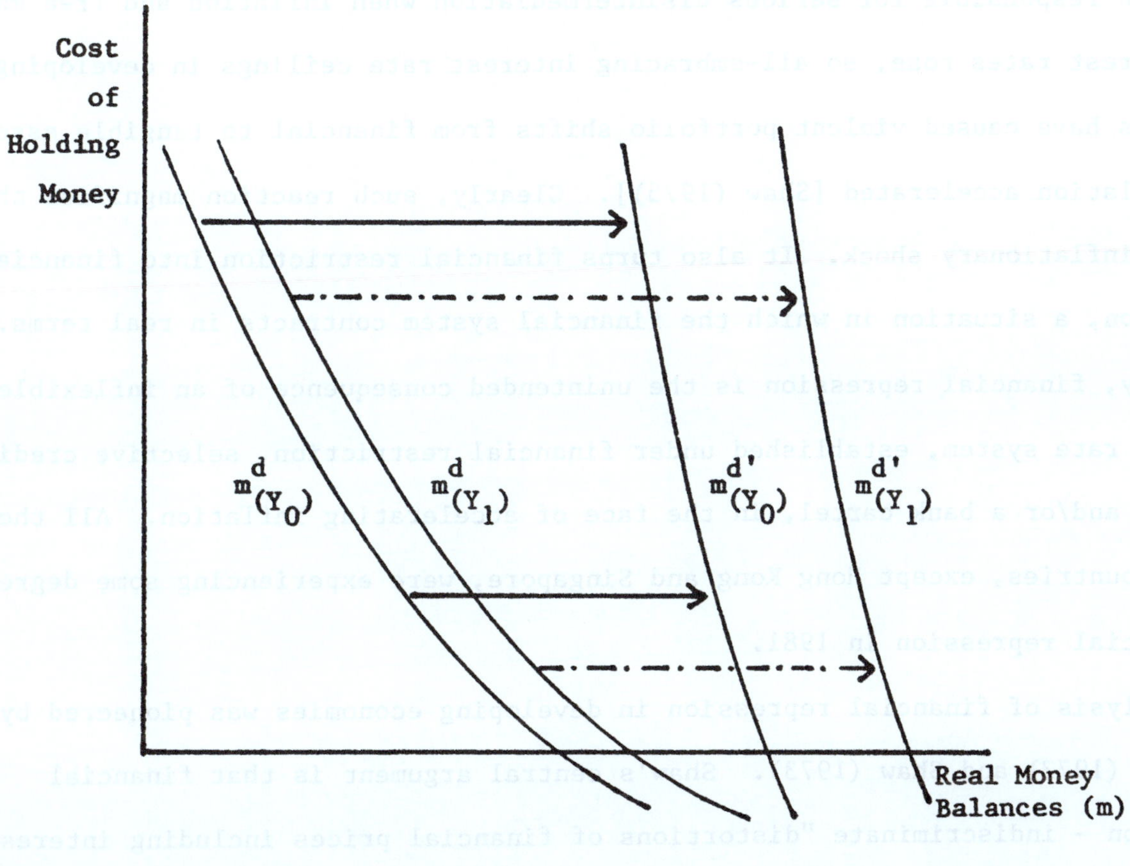

Successful financial restriction shifts the money demand function to the right at each level of income, reduces its cost-elasticity and increases its income-elasticity.

Nominal interest rate ceilings established to limit competition under policies of financial restriction can be disruptive in the face of inflationary shocks. Just as deposit rate ceilings in the U.S. and other industrial countries have been responsible for serious disintermediation when inflation and free market interest rates rose, so all-embracing interest rate ceilings in developing countries have caused violent portfolio shifts from financial to tangible assets when inflation accelerated [Shaw (1975)]. Clearly, such reaction magnifies the initial inflationary shock. It also turns financial restriction into financial repression, a situation in which the financial system contracts in real terms. Typically, financial repression is the unintended consequence of an inflexible interest rate system, established under financial restriction, selective credit policies and/or a bank cartel, in the face of accelerating inflation. All the sample countries, except Hong Kong and Singapore, were experiencing some degree of financial repression in 1981.

Analysis of financial repression in developing economies was pioneered by McKinnon (1973) and Shaw (1973). Shaw's central argument is that financial repression - indiscriminate "distortions of financial prices including interest rates and foreign-exchange rates" - reduces "the real rate of growth and the real size of the financial system relative to nonfinancial magnitudes. In all cases this strategy has stopped or gravely retarded the development process" [Shaw (1973, pp.3-4)].

The essential common elements of this model are illustrated in Figure 3. Saving, $S_{(g_0)}$, at a rate of economic growth g_0, is a function of the real rate of interest [McKinnon (1973, p.67); Shaw (1973, pp.73, 77-78)]. F represents financial repression, taken here to consist simply of an administratively

FIGURE 3

Saving and Investment under Interest Rate Ceilings

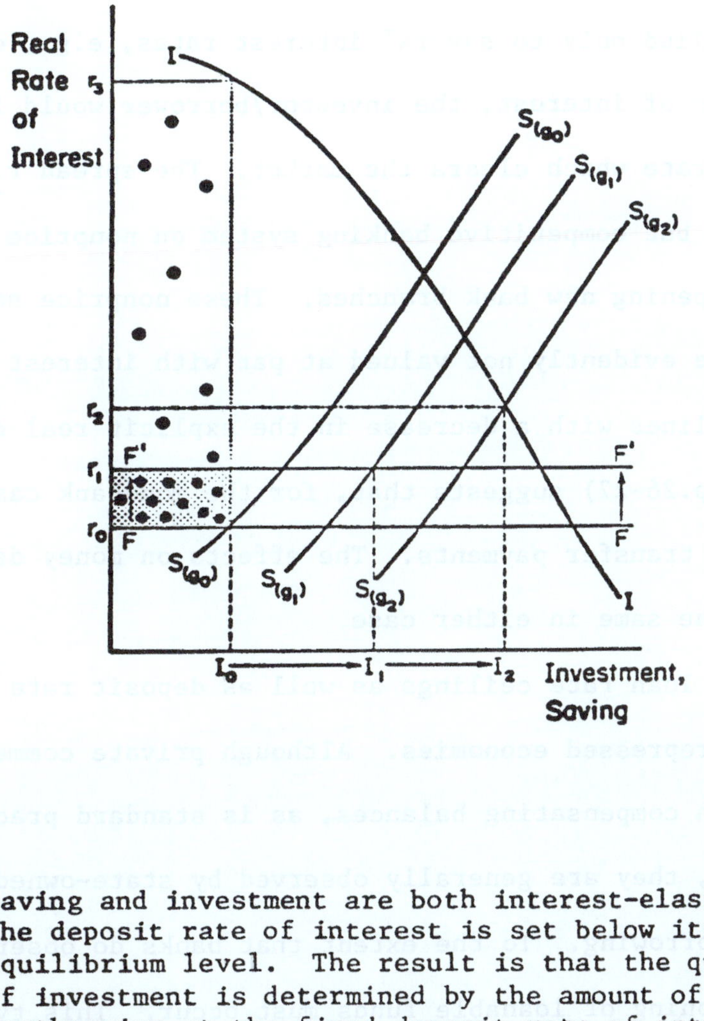

Saving and investment are both interest-elastic, but the deposit rate of interest is set below its market equilibrium level. The result is that the quantity of investment is determined by the amount of saving forthcoming at the fixed deposit rate of interest. With below-equilibrium loan rate ceilings, investment efficiency tends to be reduced too. Investments which are undertaken under these conditions are represented by the dots.

determined nominal interest rate, which holds the real rate r below its equilibrium level [McKinnon (1973, pp.71-77); Shaw (1973, pp.81-87)]. Actual investment is limited to I_0, the amount of saving forthcoming at the real interest rate r_0.

If the ceiling applied only to savers' interest rates, e.g., only to deposit but not loan rates of interest, the investor/borrower would face an interest rate of r_3, the rate which clears the market. The spread $r_3 - r_0$ would be spent by a regulated but competitive banking system on nonprice competition, e.g., advertising and opening new bank branches. These nonprice services, as already pointed out, are evidently not valued at par with interest payment - real money demand always declines with a decrease in the explicit real deposit rate of interest. Lee (1980, pp.26-27) suggests that, for the monobank case, monopoly profits are paid out as transfer payments. The effects on money demand and capital accumulation are the same in either case.

In fact, there are loan rate ceilings as well as deposit rate ceilings in almost all financially repressed economies. Although private commercial banks evade the former through compensating balances, as is standard practice in Korea and Turkey, for example, they are generally observed by state-owned banks and for all public sector borrowing. To the extent that banks do observe loan rate ceilings, nonprice rationing of loanable funds must occur. This typically takes place on the basis of quality of collateral, political pressures, "name," loan size, and covert benefits to the responsible loan officers. These criteria can be counted on to discriminate inefficiently between investment opportunities. Indeed, there will be a preference for traditional, low-yielding investments because these appear safest, simplest and cheapest to finance. Loan rate ceilings discourage risk-taking on the part of financial institutions; risk premia cannot be charged when ceilings are binding and effective. This itself rations out a large proportion of potentially high-yielding investments. There is, therefore, a strong tendency for the investments which are financed to yield returns barely above the ceiling rate r_0. These are shown in Figure 3 by the dots lying just above FF in the shaded area.

Raising the interest rate ceiling from FF to F'F', i.e., from r_0 to r_1, in Figure 3 increases saving and investment. It also rations out all those low-yielding investments, illustrated by the dots in the shaded area, which were financed before. They are no longer profitable at the higher interest rate r_1. Hence, the average efficiency of investment increases. The rate of economic growth is increased in this process and shifts the saving function to $S_{(g_1)}$.

Thus, the real rate of interest as the return to savers is the key to a higher level of investment, and as a rationing device to greater investment efficiency. The impacts on growth are multiplicative. Growth in the financially repressed economy is constrained by saving; investment opportunities abound [McKinnon (1973, pp.59-61); Shaw (1973, p.81)]. A considerable body of empirical evidence consistent with the McKinnon-Shaw model has been presented in Fry (1978d; 1978e; 1979b; 1980a; 1981a; 1981b; 1981d; 1982) and Fry and Mason (1981). In particular, investment efficiency as measured by incremental output/capital ratios was found to be positively and significantly correlated to the real deposit rate of interest. Quantitatively, economic growth seems to be reduced through lower volume and efficiency of investment by about one half of a percentage point for every percentage point by which the deposit rate is set below its competitive, free market equilibrium level [Fry (1980a; 1981b)].

Obviously, the policy prescription is to raise institutional interest rates and/or to reduce the rate of inflation. Abolishing interest rate ceilings altogether produces the optimal result of maximising investment and raising still further investment's average efficiency. This is shown in Figure 3 by the equilibrium I_2, r_2, and a higher rate of growth, g_2. Clearly, changes in the real interest rate trace out the saving function.

Lower real deposit rates of interest can be expected to reduce real money demand as well as the saving rate. Indeed, the two are connected in that a large proportion of financial saving in the sample countries is embodied in money holding [Abe *et al*. (1975); Brillembourg (1978); McKinnon (1973)]. *Ceteris paribus*, a fall in real money demand causes a decline in the real supply of credit. In practice, a fall in real money demand also reduces net foreign assets, as anticipated by the monetary approach to the balance of payments. Nevertheless, some of the effect of declining real money demand seems to be a reduction in the real supply of domestic credit.

As inflation accelerates and real deposit rates of interest fall, an increasing proportion of the declining supply of real domestic credit may be expropriated by government to finance current expenditures [Aghevli and Khan (1977 and 1978); Dutton (1971); Ness (1972); Tanzi (1977); Uluatam (1973)]. Hence, funds for both working and fixed capital investment would be doubly squeezed [Kapur (1976)].

With the real supply of domestic credit determined in the main by the real demand for money, the real deposit rate of interest determines indirectly both investment and growth. A higher real deposit rate will increase the investment rate and growth (which, in turn, feeds back to raise saving and investment rates) through the credit availability mechanism. Regression results reported in Fry

(1978a; 1978b; 1978f; 1980b; 1981a; 1981b; 1982) are all consistent with the hypotheses that the ratio of domestic credit to nominal GNP and the ratio of private sector to total domestic credit are influenced positively by the real deposit rate of interest.

One of the most deleterious effects of loan rate ceilings is that they deter risk-taking by financial intermediaries. Loan rate ceilings, when binding, eliminate the possibility of charging differential risk premia. Hence, riskier borrowers and riskier projects are rationed out completely, as illustrated in Figure 4. Here, there are two classes of borrowers/projects - less risky and riskier. A loan rate ceiling at r_0 produces a supply curve of $S_{(B_0)}$ for less risky loans and $S_{(A_0)}$ for riskier loans. The supply of less risky loans is q_0, the supply of riskier loans zero.

The abolition of loan rate ceilings shifts both supply curves. The supply of less risky loans is reduced at each rate of interest because there is some substitution into riskier loans. The supply of riskier loans at each rate of interest is also reduced, since ceiling abolition produces a higher rate for less risky lending. The equilibrium result, however, is that the actual quantities of both less risky and riskier loans are increased from q_0 to q_1 for the former, and from zero to q_2 for the latter.

There is an alternative way of looking at this effect. Consider the following example. Ten small entrepreneurs face equal value, one-year investment opportunities for which each will have to borrow 62½ per cent of the required funds. The investments will yield a 100 or -50 per cent net return with 60 and 40 per cent probabilities, respectively. The latter outcome will enable them to repay only 80 per cent of their loan principals and no accrued interest.

FIGURE 4

Riskier and Less Risky Loan Markets

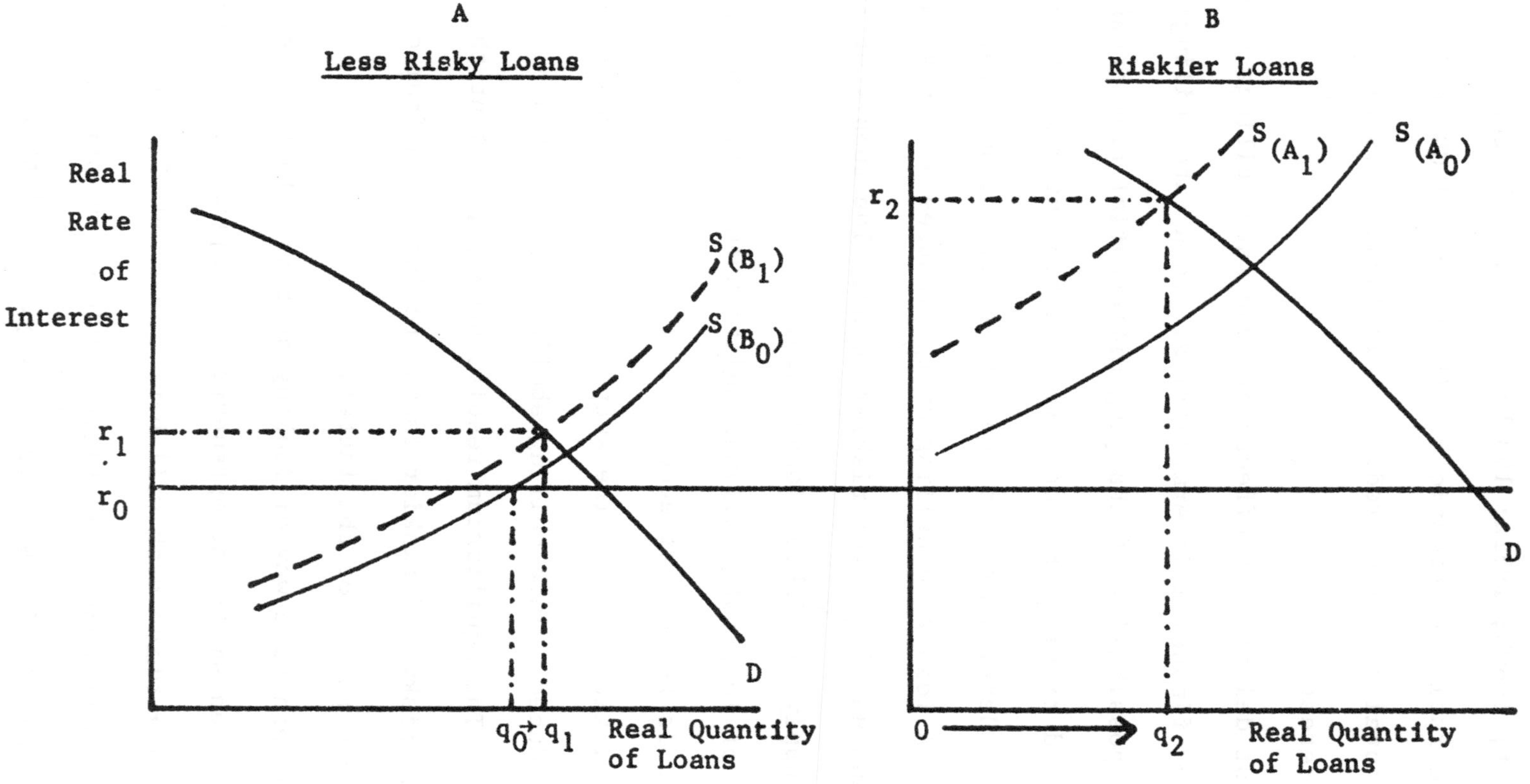

Panel A illustrates the market for less risky loans, panel B the market for riskier loans. A uniform loan rate ceiling at r_0 rations out completely all riskier borrowers - supply of riskier loans is zero at r_0. Abolishing the loan rate ceiling produces an equilibrium interest rate for less risky loans of r_1 and for riskier loans of r_2. The quantity of riskier loans supplied rises from zero to q_2.

Assume that the lender is risk-neutral and attempts to maximise profits. If he lends Rs 1 million at 30 per cent interest to each of these entrepreneurs, the expected net return is exactly 10 per cent; 30 per cent on Rs 6 million and -20 per cent on Rs 4 million. Hence, the lender would turn down all 10 loan applications if the loan rate ceiling were below 30 per cent, provided the alternative was riskless assets yielding 10 per cent. The critical loan rate ceiling is 13-1/3 per cent if the alternative were idle cash yielding nothing. Despite the fact that the 10 investment projects yield an average economic return of 40 per cent, i.e., in total are highly productive, they might all be rationed out with an interest rate ceiling of 13 per cent. A profit-maximising lender would rationally prefer to hold idle cash than to make these risky loans under such circumstances.

In fact, nationalised financial intermediaries are often directed to take risks and to extend small loans which are more expensive than larger loans to administer without regard to compensating returns. For example, the overall cost - including expected delinquency and default costs - of providing credit to small farmers in India averages 2¼ per cent more than the cost of providing credit to large farmers. Consequently, smaller farmers are subsidised to a greater extent than larger ones, even with a uniform loan rate.

The result of disregarding risk has been serious levels of delinquency and default in the loan portfolios of many public sector financial intermediaries. For example, about 50 per cent of India's land development banks' loans are delinquent. When the distinction between a loan and gift becomes blurred, collection of loans by other financial intermediaries is affected. Private money lending in rural areas of India has declined dramatically in recent years, due in part to increasing collection difficulties. Similar problems of unacceptably high delinquency and default rates on priority loans extended by government-owned financial intermediaries exist elsewhere, e.g., in Indonesia, Korea, Nepal and Western Samoa. Half of the Development Bank of Western Samoa's agricultural loans have had to be rescheduled.

Ultimately, the costs of high delinquency and default rates are borne by depositors. Hence, they reduce the aggregate real supply of loanable funds. At the same time, delinquency and default rates of the magnitudes found, for example, in India, Indonesia, Korea, Nepal and Western Samoa reflect adversely on both the administrative and allocative efficiency of the public sector financial intermediaries. Of course, risks are an inherent part of the process of economic development. But performance criteria should promote profitable, productive risk-taking and deter mere indiscriminate lending.

Abolition of loan rate ceilings can increase the average efficiency of investment, because higher rates will reduce the demand for investible funds by those with relatively low-yielding investment projects. This frees resources for use by those who were previously rationed out of the market but who do have projects, albeit risky, with high expected returns. An increase in the incremental output/capital ratio and, hence, in the rate of economic growth can be anticipated.

Abolishing interest rate ceilings can stimulate competition not only among financial intermediaries but also between them and the bond markets, which can then be developed more effectively. Increased competition will help small and medium size enterprises in particular, because they are the first to be rationed out under noncompetitive conditions. With ceilings, bankers can live an easy, quiet life. They wait for customers, waste resources on elegant buildings, and turn down applicants without "name," first-class credit standing, etc. Initiative and new enterprise are suffocated. Since new ventures are invariably risky, they tend to be doubly discriminated against by the inability of the banks to charge any risk premium when loan rate ceilings are effective.

Interest rates can perform three basic functions [Chandavarkar (1971, p.50)]. First, the interest rate can mobilise saving. It is the price which influences the choice between present and future consumption. Under the imperfect market conditions found in almost all the sample countries, it can also be expected to have a strong effect on the choice of assets in which savings are embodied. A rise in institutional interest rates produces substitution from unproductive tangible assets held as inflation hedges into financial claims. This substitution as well as any increase in the saving rate frees resources for productive investment.

Second, the interest rate is an efficient rationing device for the allocation of scarce resources between alternative investments. It is almost invariably superior in this respect to rationing on the basis of decisions of a bureaucrat, quality of collateral offered, political influence of the borrowers, "name," or the personal preferences of individual loan officers. As a rationing device, the interest rate maximises the average return of a given volume of investment.

Third, the interest rate can provide a social discount rate for decisions both to save and to invest. In this role, it equates planned saving and investment. Here it acts as a market clearing device, influencing in an optimal manner the choices of what to produce and how to produce it. The interest rate can discourage the establishment of automobile or aircraft factories, economic activities which are inherently capital intensive. The interest rate can also discourage capital intensive techniques of production for a particular product in countries where capital is limited. Where labour is plentiful and capital is scarce, the interest rate can direct entrepreneurial activities into simple things with simple technologies, but with high returns to capital.

The interest rate performs these three functions automatically when it is allowed, as it is in Singapore, to find its equilibrium market level through the free competitive interaction of the forces of supply and demand. It performs none of these functions effectively in most of the other sample countries because ceilings are imposed by fiat or cartel agreements on so many institutional interest rates. One suspects strongly that the low interest rate policy pursued by these countries, particularly since the 1960s, has reduced saving and investment and lowered the average efficiency of the investment which has taken place.

The optimal solution to loan rates is the abolition of ceilings and the abandonment of selective credit policies. If certain economic sectors are to be subsidised, subsidies could be given in fiscal form rather than through low interest rates. Ministries of finance object that this raises Budget expenditures. In fact, however, the cost of the fiscal subsidy can be identical to that of the interest rate subsidy. The tax which financed the interest rate subsidy, i.e., deposit rates held below their market equilibrium levels, can finance a direct fiscal subsidy or grant instead. Preferably, the tax system would be reformed at the same time so that the implicit deposit tax - a socially inefficient tax on saving - is dropped in favour of another tax which is not so inefficient. Fiscal subsidy has two other advantages: it can be confined more easily to the priority activity itself and it need not distort factor prices. Labour as well as capital can be subsidised.

5. Foreign Financial Intermediaries in Small Island Developing Economies

A bird's eye view of the financial sectors in all the sample countries is presented in Table 3. Of particular note is the fact that only Hong Kong and Singapore have any private joint stock domestic commercial banks. Barbados, Fiji and Papua New Guinea each have one government-owned domestic commercial bank and the governments of the Solomon Islands and Western Samoa have both participated with a foreign commercial bank in the establishment of one domestic commercial bank. A government-sponsored cooperative bank exists in St. Lucia. In three of the sample countries - Bahamas, the Maldives and Seychelles - there are no domestic commercial banks and all banking business is undertaken by branches of foreign commercial banks.

Commercial banks dominate the financial sectors of all the sample countries, except Hong Kong and, to a lesser extent, Singapore. Bahamas and Papua New Guinea have a few savings and loan associations (S&Ls), the Seychelles has the Government Savings Bank and there is a Post Office Savings Bank in Western Samoa. These are the only nonbank depository institutions outside Hong Kong and Singapore in all the sample countries. Other nonbank financial intermediaries found in the sample countries include pension/provident funds, insurance companies, finance companies (depository institutions found only in Hong Kong), trust companies and stock exchanges (in Fiji, Hong Kong and Singapore). There are also government-owned development banks and/or similar developmental financial intermediaries in all the sample countries, except Hong Kong and the Maldives. Finally, five countries - Bahamas, Barbados, Hong Kong, Seychelles and Singapore - have offshore banking. The operations in Barbados and Seychelles are still at an embryonic stage. The big offshore centres are located in Bahamas, Hong Kong and Singapore.

TABLE 3

Bird's Eye View of Financial Sectors in Sample Countries

Country	Monetary Authority	Commercial Banks Domestic	Commercial Banks Foreign	Nonbank Financial Intermediaries
Bahamas	Central Bank	0	11	4 S&Ls; 9 trust companies.
Barbados	Central Bank	1 (government-owned, universal bank)	5	Trust companies.
Fiji	Central Monetary Authority	1 (government-owned)	5	National Provident Fund; insurance companies; Unit Trust of Fiji.
Hong Kong	Financial Secretary	32 (all private)	81	265 deposit taking companies; insurance companies; money and commodity brokers.
		(universal banks)		
Maldives	Department of Finance	0	2	None
Papua New Guinea	Central Bank	1 (government-owned)	3	S&L societies; 1 merchant bank.
St. Lucia	Member of East Caribbean Currency Authority	1 (cooperative bank)	4	None
Seychelles	Monetary Authority	0	6	Government Savings Bank.
Singapore	Monetary Authority and Board of Commissioners of Currency	12 (all private)	37	Central Provident Fund; Post Office Savings Bank; insurance companies.
Solomon Islands	Monetary Authority	½ (government-owned)	2½	National Provident Fund.
Western Samoa	Monetary Board	½ (government-owned)	1½	National Provident Fund; Post Office Savings Bank; Public Trust Office; insurance companies.

TABLE 3 (Continued)

Country	Government-owned Development Banks	Institutional Interest Rate Setting Procedure	Offshore Banks	Remarks
Bahamas	0	No formal controls but bank cartel fixes rates in consulation with central bank.	263	
Barbados	1	Minimum deposit rates and maximum weighted-average loan rate set by central bank.	1	Some private companies and individuals borrow and lend directly without going through financial intermediaries because of wide spread.
Fiji	1	Maximum deposit and loan rates established with approval of Ministry of Finance.	0	Stock exchange established in 1979. Eight companies listed by June 1980.
Hong Kong	0	Maximum deposit rates fixed by bank cartel led by Hong Kong and Shanghai Bank.	378	Hong Kong and Shanghai Bank performs several central bank functions. Currency notes are issued by commercial banks which are universal banks in character.
Maldives	0	No controls. Loan rate is based on New York prime plus 1½ per cent. Bank cartel suspected.	0	Maldives Monetary Authority is due to be set up.
Papua New Guinea	2	Rediscount rate and moral suasion used by central bank to influence deposit and loan rates. Bank cartel suspected.	0	
St. Lucia	4	No control. Bank cartel suspected.	0	
Seychelles	1	Prime rate fixed by Monetary Authority. Maxima and minima for deposit and loan rates set by bank cartel in consultation with Monetary Authority.	1	
Singapore	1	All interest rates are determined competitively in a free market. Bank cartel was abolished in 1975.	70	
Solomon Islands	2	Monetary Authority fixes deposit rates and minimum and maximum loan rates.	0	The post office sells 3-year government national savings certificates. 5-year government development bonds are sold through Monetary Authority and commercial banks.
Western Samoa	1	Government fixes all institutional interest rates. Highly selective credit policy is pursued.	0	The Bank of Western Samoa is owned jointly by the government and a New Zealand commercial bank.

As pointed out earlier, there is strong correlation between strength and breadth of these financial sectors and economy size. The financial sector of the Maldives with a 1979 GNP of $30 million is rudimentary in the extreme. Financial sectors of St. Lucia ($95 million GNP), Seychelles ($91 million), Solomon Islands ($96 million) and Western Samoa ($66 million) are just one rung further up the ladder. Papua New Guinea ($1,950 million GNP) has a marginally more extensive financial sector, despite the fact that its per capita income is about half that of Seychelles. The "big" economies of Hong Kong ($19 billion GNP) and Singapore ($9 billion) possess financial sectors comparable to any found in the Western industrial countries. The correlation is caused, in the main, by the economies of scale which exist in financial intermediation.

In unit banking states in the U.S., there was one bank for every $65 million of 1979 GNP. However, these banks can and do share services such as computer facilities, etc., so enabling them to reap the benefits of scale economies. Furthermore, they can easily purchase intermediate inputs such as cheque books from specialised firms supplying the U.S. banking industry as a whole. Labour costs of all insured banks in the U.S. were 1.6 per cent of their earning assets in 1976. The comparable 1977 figure for Turkey, for example, was 5.8 per cent. Unit labour costs of banking in the U.S. cannot, therefore, be presumed to exceed unit labour costs of banking in any of the sample countries.

From all this, one might accept that the minimum economy size needed to support one viable de novo domestic bank would be a 1979 GNP of at least $100 million This implies that not even one private domestic bank could survive without government subsidy in the Maldives, St. Lucia, Seychelles, the Solomon Islands and Western Samoa. On this estimate, Bahamas and Barbados could support six each, Fiji ten and Papua New Guinea 20 private domestic banks, provided each bank had only one branch.

The estimate for Papua New Guinea brings home the fact that not only is economy size crucial but so also is expertise. Qualified personnel to run 20 private domestic banks in Papua New Guinea is just not available there. Even with adequate economy sizes, Afghanistan and Nepal illustrate well the problems of establishing and maintaining all-domestic financial sectors without a sufficient supply of trained personnel. Specifically, foreign trade financing constituted a serious impediment to foreign trade in these countries. The banks could not be relied upon to execute correctly such standard operations as opening letters of credit [Fry (1974a, Chs. 5 and 8; 1974b; 1976b, p.1137; 1978e)]. Foreign banks tend to have strong comparative advantages in terms of expertise and experience in foreign trade finance. For some of the sample countries, economy size, training and know-how deficiencies leave no room for choice. Either financial intermediation is undertaken by foreign banks or there will be none at all.

Grubel (1977) has provided probably the most comprehensive framework with which to assess the benefits and costs of permitting foreign banks to establish branches in countries where a choice is actually possible. The main advantages to be measured are: (a) increased competition forced on domestic financial intermediaries which would form an oligopolistic if not cartelised industry in the absence of foreign banks; (b) the importation and use of existing stocks of knowledge capital or know-how at a very low marginal cost; and (c) increased efficiency of international capital flows [Grubel (1977, pp.357-358)].

The disadvantages of hosting foreign banks, according to Grubel, may include: (a) their exemption from socially beneficial regulations; (b) loss of control over eurodollar liquidity; and (c) inflation created by multiple deposit creation of eurodollars [Grubel (1977, p.358)]. Clearly, these disadvantages are valid, if they are at all, only for offshore foreign banking.

For the present purposes, the potential disadvantages which should be considered include: (a) inefficiency in resource mobilisation and allocation resulting from unfamiliarity with local conditions; (b) export of national saving to country of origin resulting again from unfamiliarity with or unresponsiveness to local conditions; (c) formation of a foreign bank cartel which dictates or thwarts monetary policy measures; and (d) increased difficulty of starting a domestic bank when foreign banks are well established and hold all the best accounts.

The four potential disadvantages listed above can be realised and magnified all too easily by inappropriate laws and regulations. Inefficient resource mobilisation and allocation is assured when the monetary authority sets binding deposit and loan rate ceilings. Conversely, permitting and encouraging mobile bank branches, providing partial loan guarantees, ensuring prompt and effective legal redress in cases of loan delinquencies and defaults, contributing towards training costs, etc., can promote efficient resource mobilisation and allocation on the part of both domestic and foreign financial intermediaries.

Charging licence fees, as St. Lucia does for both head offices (EC$20,000 annually) and branches (EC$1,000 each), is likely to reduce efficiency of financial intermediation. Setting minimum deposit rates, as Barbados does, may raise efficiency. No matter how small the number of financial intermediaries, whether domestic or foreign, regulations can be desfgned and implemented to simulate competitive conditions, i.e., by setting minimum deposit rates of interest. Equally important, discriminatory taxation of financial intermediation can and should be avoided in the interests of efficient financial intermediation.

Most measures discussed above in connection with increasing or decreasing financial intermediation efficiency also affect the incentive on the part of foreign financial intermediaries to siphon off national saving to their home countries. Additionally, a rapidly depreciating domestic currency tends to stimulate capital flight, whereas a smoothly appreciating exchange rate may well deter it.

That foreign bank cartels in the sample countries can indeed influence if not dictate domestic monetary and/or fiscal policy is illustrated by a recent event in St. Lucia. In 1980, the government proposed, albeit misguidedly, to impose a 2 per cent tax on banks' deposit liabilities. The proposal was dropped after discussions with the foreign banks. A potential or existing foreign bank cartel may perhaps best be averted or undermined by relatively free entry conditions. However, very small island developing economies could realise substantial savings in information gathering costs from granting licences only to large, reputable foreign banks which would be least likely to risk adverse publicity from sharp, if not illegal, practice. There might also be some advantages, as Seychelles appears to have recognised since 1976, in encouraging foreign banks from several different countries to set up offices.

There seems to be no proven way of solving the problem of subsequent domestic entry into the financial intermediation industry which historically has been the exclusive territory of foreign financial intermediaries. Government-ownership is the only method so far attempted with respect to commercial banking in all the sample countries, except Hong Kong and Singapore. An alternative technique could be to reserve the field of thrift intermediation, i.e., S&Ls, finance companies, trust companies and mutual savings banks, for domestic enterprise. At an appropriate moment, thrifts could be put on an equal footing with commercial banks, as is now happening in the U.S. Yet another alternative might be to issue fixed period, e.g., 15-year, licences to foreign banks with an agreement that the intermediary would be transferred to national ownership at the expiration of the licence.

This section concludes with a brief discussion of offshore banking in the sample countries. Offshore banking consists of banking operations conducted solely for and with nonresidents in foreign currency denominated claims. Offshore centres can be classified as "paper" or "functional" centres [McCarthy (1979, p.45)]. Paper centres act solely as locations of record for the purpose of tax avoidance. Functional centres actually carry out deposit taking and lending, acting as "important links between Eurocurrency markets, helping to channel funds from major international financial centers (such as London and New York) to final borrowers" [McCarthy (1979, p.45)].

Three of the sample countries - Bahamas, Hong Kong and Singapore - host large functional offshore banking centres. Recently two others - Barbados and Seychelles - enacted the legislation necessary to permit offshore banking. Both of these countries have so far each attracted one offshore bank. Several other countries in the sample are flirting with the idea of opening up offshore centres. It is, therefore, appropriate to consider here some of the major benefits and costs of hosting offshore banks.

The direct benefits to be derived from offshore banking centres include: (a) licence fees and profit taxes; (b) domestic currency capital and liquidity requirements; and (c) the benefit of the offshore banks' operating expenditures. The indirect benefits may consist of: (a) improved access to international capital markets; (b) increased efficiency of domestic financial intermediaries resulting from greater competition and/or demonstration of more sophisticated banking techniques; (c) training; and (d) attraction of ancillary institutions, e.g., insurance, merchant banking, brokerage houses, etc.

Few of the benefits listed above can be expected from paper offshore centres. Indeed, paper centres yield only extremely modest income from licence fees and taxes. The demand for such centres springs from a desire to avoid taxes. There is a competitive supply of paper centres. Hence, licence fees of a few thousand dollars are sufficient to divert business elsewhere. Paper centres incur virtually no local operating expenses, bar the costs of cleaning the brass plates. And, clearly, there can be no indirect benefits.

In functional centres, licence fees and profit taxes levied on offshore banks can produce significant revenue. In the Cayman Islands, for example, licence fees paid by offshore banks produce annual revenue equal to $70 per capita. Withholding tax on interest earnings appears to inhibit offshore banking, as evinced by Hong Kong's experience. Profit taxes, on the other hand, are clearly not so much of a deterrent, as the Singapore case attests. And Bahamas has raised revenue successfully through stamp duties.

Domestic capital and/or reserve requirements - capital requirements of this type are imposed by Panama and Singapore - yield a net benefit equal to the return on the funds expropriated minus any interest earned by the offshore banks on their local capital or reserves. Such requirements have not commonly been imposed. Like withholding taxes on interest earnings, they tend to deter offshore banking.

Operating expenses are probably the most important benefit accruing from functional offshore centres. Of course, the net benefit is the relevant figure. This can be calculated by subtracting expatriate labour costs, imports' costs and the opportunity cost of local employees and local capital, e.g., buildings, from the gross expenditures of the centre. However, indirect taxes, e.g., import duties, should be added back to the net benefit figure. Indeed, import duties paid by offshore banks can be considerable.

The indirect benefits of offshore banking centres are much more difficult to measure. The supposed improved access to international capital markets is likely to be due to the same factors that attracted the functional offshore centre in the first place, rather than to the subsequent existence of the centre itself. Competition will not be stimulated if offshore banking is kept, as it easily can, strictly separated from the onshore financial sector. Training may be of little benefit to onshore financial intermediaries if offshore banks keep the trained labour. Indeed, they may create a brain drain. Finally, only the largest functional offshore banking centres can expect to attract any significant number of ancillary institutions. In sum, the net benefits may not be as substantial as has sometimes been thought [McCarthy (1979, p.48)]. Hodjera (1978, p.242) estimates that Singapore's large functional offshore banking centre contributed only about 1 per cent of gross domestic product in 1976.

On the obverse side of the coin are the costs involved in attracting offshore banks. The direct costs include: (a) legislative and regulatory changes; (b) communications; (c) education and training; and (d) regulation and supervision. There may possibly also be an indirect cost in the form of weakened monetary autonomy.

The largest costs of attracting offshore banking involve the expenses of installing and maintaining a modern and highly efficient communications system. Functional offshore centres will not be attracted unless the local population has received a reasonable level of education and training.

Regulation and supervision can be extremely costly. The results of easy entry combined with the absence of any supervision are well illustrated by the recent case of St. Vincent. Towards the end of 1978, St. Vincent opened its doors to all foreign banks wanting to establish offshore branches. No regulations or supervision were imposed. By early 1979, organised crime had become well established in St. Vincent's offshore banking centre. After the U.S. exerted pressure, St. Vincent closed the centre down amid a considerable scandal in mid-1979. Because of the high costs of supervision, it may well pay to licence only large, reputable banks and dispense with formal supervision on the grounds that such banks will not risk any activities which might cause scandal.

McCarthy (1979, p.48) concludes his analysis of the benefits and costs of hosting offshore banking centres as follows:

> While the benefit-cost equation appears favorable for existing centers, it seems possible, indeed probable, that there is little unsatisfied demand for new offshore centers. There are even some signs at present of an excess supply. If one looks at the existing geographical coverage provided by offshore centers, virtually every area of the world has a selection of offshore centers readily accessible. In addition, improved telecommunications render it easier than before to route paper business through a limited number of centers rather than setting up operations in several widely dispersed centers. In general, therefore, new paper centers are not likely to succeed. In addition, even existing centers might become less important and less profitable if moves to impose controls on the Euro-currency markets are successful, and/or if New York establishes an International Banking Zone.

The probability of New York's International Banking Zone getting launched was increased considerably at the end of 1980 by the submission of the Federal Reserve Board's proposal to permit U.S. banks to establish International Banking Facilities within the U.S. [Cheng (1981)]. The acceptance of this proposal would be a crippling blow to some of the existing offshore banking centres, particularly those located in the Caribbean.

6. Policies for Effective and Efficient Domestic Resource Mobilisation and Allocation in Small Island Developing Economies

The objective of any financial development programme is to raise both the quantity and quality of investment and, hence, to accelerate the rate of economic growth. The majority of such programmes have, in practice, stressed institution building plans - development banks, stock exchanges, etc. They incur substantial resource costs. However, Khatkhate and Villanueva (1978, p.982) conclude:

> The evidence is strong that the specialized institutions are no panacea for solving the basic problem of credit allocation ....

An appropriate legal framework together with price stability are two basic prerequisites for promoting efficient financial intermediation. The legal framework will determine, in large part, the structure of the financial sector. Khatkhate and Riechel (1980) point out the drawbacks of banking laws which enforce specialisation:

> In developing countries, demand for even basic financial services has often not yet been appropriately articulated. In such situations, it appears desirable to generate through official intervention such special sources of supply that can meet socially desirable, albeit partially dormant, private demand. For this purpose, developing countries have often established new specialized financial institutions to satisfy the previously unmet demand. Operations of such institutions are generally insulated from competition by appropriate legislation and are even given substantial subsidies. Such actions are often defended by arguments that resemble those employed in the infant industry advocacy. However, the efficiency gains expected from such specialized and protected institutions are unlikely to be realized, because the necessary competitive conditions are often absent. In fact, a specialized institution created by special statute often assumes a monopoly position. The establishment of a special institution can be justified only if it will expand the overall size of the financial sector, widen its spectrum of financial services, and reduce the degree of concentration. In order to accomplish these goals, the new institution needs to be broadly based and, after the infancy phase is over, needs to be exposed to competitive forces across the board.
> [Khatkhate and Riechel (1980, pp.504-505)]

> The foregoing analysis suggests that the fragmentation of the financial sector that follows from legislated specialization tends to produce two undesirable consequences: a decline in overall efficiency and an increase in the degree of concentration.
> [Khatkhate and Riechel (1980, p.502)]

Khatkhate and Riechel advocate multipurpose or universal banking, citing the German experience as an object lesson for most developing economies, in particular small ones [Khatkhate and Riechel (1980, pp.481-487)]. Legislative and regulatory changes are required in most of the sample countries, with the notable exception of Hong Kong which already has a universal banking system, to permit multipurpose banking.

Four benefits can be derived by switching from specialised to universal banking: (a) improved economic efficiency; (b) more long-term capital; (c) promotion of entrepreneurship; and (d) greater financial stability. Efficiency can be raised through adoption of a universal banking system which can reap greater economies of scale, be more responsive to changing demands and, at the same time, exhibit increased competitiveness and reduced concentration [Khatkhate and Riechel (1980, p.493)].

Universal banks can supply both short and long-term capital from resources mobilised directly from savers. Specialised development banks - the only suppliers of long-term funds under a specialised system - invariably get most of their resources from other financial intermediaries, e.g., the central bank. Not only does this create inefficient financial layering, it also results almost always in an acute shortage of long-term investible funds.

Universal banks can be expected and encouraged to promote entrepreneurship, as they did during Germany's industrialisation, by offering packages of credit combined with managerial and technical assistance, much as the World Bank does today.

Finally, multipurpose banks may well be less prone to financial instability than specialised financial intermediaries. Scale economies enable them to acquire more information cheaper, transform more short-term liabilities into longer-term assets and diversify their portfolios to greater extents than can smaller, specialised banks [Khatkhate and Riechel (1980, pp.495-497)].

A move towards universal banking has already started in the Western industrial countries, as well as in a few developing countries:

> .... many other developing countries .... sorely disappointed by a lack of long-term financing for development programmes .... are following the richer world's example. They are going down the path of universal banking.
> [Economist (1981, p.96)]

The major potential danger of universal banking, a system under which financial intermediaries are permitted to acquire equity interests in their borrowers, lies in the conflict-of-interest issue. However, appropriate legislation can confront this potential problem in advance:

> .... conflict-of-interest situations and the prevalence of excessive market power require legislation, like antitrust laws, oriented toward tackling these problems directly rather than indirectly through a narrowing of the range of activities a financial institution can cover. The possibility that such unsavory practices may recur should not be taken as a pretext to devise a straightjacket of banking legislation that would destroy the responsiveness, flexibility, versatility, and the dynamism of the financial system.
> [Khatkhate and Riechel (1980, p.513)]

The case for a legislative framework conducive to the development of universal banking is strong. For small island developing economies, the case for encouraging, without discriminatory taxes, large, reputable foreign banks to help develop a universal banking system is even stronger.

Financial development programmes have exhibited a high propensity to collapse in the face of accelerating and increasingly volatile inflation. McKinnon (1973, pp.77-79) and Shaw (1973, pp.119-120) both emphasise low and stable inflation as part of and a prerequisite for financial development. Hence, the foundation stone of any financial development programme in a small island developing economy is control over domestic credit expansion to ensure that an exchange rate policy consistent with low and stable rates of inflation can be pursued. Governments would first have to decide that they are willing to accept the fiscal discipline moderate credit expansion imposes. Then they might give authority and responsibility to the central banks for keeping domestic credit expansion on its noninflationary target.

Financial development and innovation can start once basic financial reform has taken place. Figure 5 shows the objective of financial reform and financial

FIGURE 5

Effects of Financial Repression, Financial Restriction, Financial Reform and Financial Development on Saving and Investment

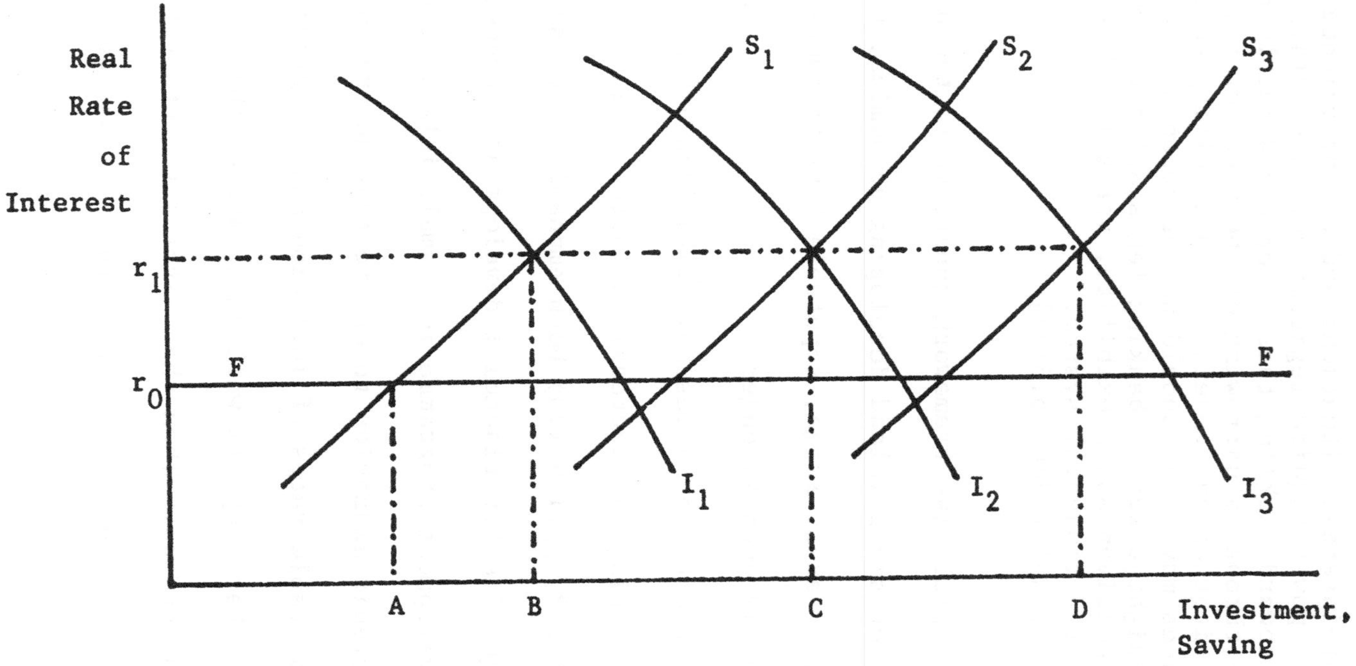

The saving and investment functions shift upwards and to the right with each step in the programme of financial development. This is achieved by measures which reduce the spread between savers' net returns and investors' gross costs of borrowed funds.

development. Point A is the level of saving and investment under financial repression. The negative real interest rate caused by the ceiling reduces saving to A. Investment can be no greater than foreign saving plus the level of national saving forthcoming at the real interest rate r_0. Low average investment efficiency is anticipated, despite the low quantity of investment.

Point B represents the situation under financial restriction. The interest rate r_1 is the "market" rate, but net returns to savers are well below r_1 and gross costs to borrowers are considerably above r_1. High reserve requirements, taxes on financial transactions and on income from savings, deliberate suppression of certain financial markets, etc., all widen the gap between the net return to savers and the gross cost to borrowers, even in the absence of interest rate ceilings. With interest rate ceilings as well, financial restriction produces a level of saving somewhere between A and B along S_1. This, of course, will also be the level of investment.

Point C is the level of saving and investment achieved after a basic financial reform. Interest rate ceilings have been abolished and the government has accepted the proposition that cheap public finance is likely to impose a high social cost. The government would reduce its seigniorage from the money supply and instead start offering positive real returns on government bonds. Maybe too, some measures are taken to reduce taxes on saving and investment.

Point D is reached only after a fully comprehensive programme of financial development to raise both the quantity and the average efficiency of investment has been implemented. There is now a considerable body of evidence indicating that direct fiscal incentives have negligible or even perverse effects on the level of investment and actually reduce average efficiency [Nickell (1977);

Tanzi (1976); Usher (1977)]. However, financial development implemented in conjunction with liberalisation of fiscal, price and foreign exchange rate policies can itself stimulate investment.

Fiscal incentives have been employed in a number of developing countries not only to encourage investment but also to raise saving propensities. As with investment incentives of this kind, it is not clear that piecemeal fiscal measures to raise saving rates have, in general, been effective. Nevertheless, there is evidence suggesting that saving is sensitive to the net real return on financial assets, as documented above. Income tax lowers the net real return to saving, so reducing saving and investment rates. This, in turn, reduces the capital/labour ratio and labour productivity. Were income from capital not taxed at all, saving would rise and the return to capital would consequently fall. Capital's gross share in GNP could fall substantially. The consumption tax may, in fact, be less regressive than an income tax in the long run. The adoption of a consumption tax could help shift the saving function from S_2 to S_3 in Figure 5 [Boskin (1978)].

An important item in this financial development programme involves paying a competitive return on banks' required reserves. In principle, the reserve ratio can be set at any level - even zero - for purposes of monetary control. However, the magnitude of that ratio should be decided with the objective of parcelling out the lending function efficiently between the central bank and the deposit institutions [Fry (1979a)]. Reserve requirements drive a wedge between returns to savers and costs to borrowers. They lower rewards to savers and raise rates paid by investors, thereby reducing incentives both to save and to invest. Reserve requirements may be viewed as an inefficient tax for this reason, as already pointed out.

Required reserves might earn, say, 85 per cent of the normal short-term loan rate. By exactly how much the rate on required reserves should be below the loan rate involves detailed cost calculations beyond the scope of the present paper. Here, the intention has been simply to explore the concept of a competitive interest payment on required reserves. The reserve requirement tax is reduced to zero if the reserve deposit rate of interest is precisely equal to equivalent commercial bank loan rates less the cost that financial institutions incur in lending and servicing loans.

Who benefits from the refund depends, of course, on the incidence of the reserve requirement tax. Initially, all gains are at the expense of taxpayers. The classic story about this tax refund is that it accrues substantially to depositors who respond by increasing their real demand for deposits. Then economies of scale reduce the unit operating costs of financial intermediation and loan rates to investors can decline. The final outcome is more rapid capital formation and a higher rate of economic growth. Indirectly, all taxpayers benefit. Putting aside implications for the Budget, the optimal monetary policy here is a uniform reserve requirement and a fully compensatory yield on required reserve balances.

A competitive banking system is imperative for financial development. Setting minimum deposit rates of interest has already been suggested as a device which might be used to force uncompetitive banking systems to seek out borrowers in a competitive manner. At the same time, interest rate flexibility has been advocated to prevent falling money demand for exacerbating inflationary pressures created, for example, by monetary accommodation of exogenous supply shocks.

The two proposals could be combined through the introduction of an indexed negotiable certificate of deposit (NCD) of, say, five-year maturity. Banks might be obliged to offer these NCDs, whose principals would be indexed. At redemption, the face value of the NCD would be adjusted by the cumulative change in the price index adopted. Annual interest payments of, say, 3 per cent would also be adjusted by the same index.

These NCDs may serve three purposes. First, they lengthen the maturity of banks' liability portfolio, so enabling the banks to increase their medium and long-term lending activities without undue risk. Second, trading in NCDs produces a simple market in which experience with a market-determined yield can be gained with minimum risk.

The third function of the NCDs is to stabilise aggregate real money demand in the face of volatile inflation. To understand this point, one might assume that savings can be distributed among four assets - currency in circulation, C, demand deposits, D, negotiable certificates of deposit, T, and unproductive tangible assets held as inflation hedges, A. If C, D, T and A possess just two attributes - return and liquidity - the ranking by return must necessarily be identical to the ranking by illiquidity. In this case, it can be shown that substitution will take place only between adjacent assets [Barrett, Gray and Parkin (1975, pp.505-508)]. This means that the demand for A is determined only by r_A, the real return on A, and r_T, the real return on T. Hence, aggregate real money demand too is a function only of r_A and r_T. Returns on C and D will affect the composition of the money stock, but not the aggregate real demand for money [Fry (1978c)]. Evidently, real money demand can be stabilised by holding just the difference between r_A and r_T constant. This is exactly what the untaxed indexation of NCDs or time deposits achieves [Bhatia (1974)].

Deposit indexation may be a radical reform because it breaches, where they exist, nominal interest rate ceilings. The main benefits of deposit indexation imposed on a noncompetitive banking system are: (a) it stabilises real money demand [Simkin (1978, p.130)]; (b) it forces financial intermediaries to behave competitively in their search for borrowers; and (c) it enables long-term lending to continue under conditions of volatile inflation [Bhatia (1974)].

Loan indexation might also be considered, particularly in high inflation countries. The point of indexing medium and long-term loans is to lengthen their effective maturities. High inflation and concomitant high nominal loan rates reduce effective maturities by requiring borrowers to accelerate repayment of principal. The high nominal loan rates simply compensate for the fall in the real value of the principal outstanding. Hence, they can be viewed in real terms as accelerated principal repayments [Fry (1980c)].

Indexing all deposits or NCDs or time deposits only may cause substantial fluctuations in the currency/deposit ratio, since the difference between the return on currency and the return on deposits will vary with variations in the inflation rate. Monetary control through the cash base can be jeopardised, particularly in developing countries suffering long lags in compilation of monetary statistics. It is possible, however, to stabilise the money multiplier through reserve requirement policy [Fry (1979a)].

The indexed NCDs might be the first step towards the complete abolition, where they exist, of interest rate ceilings. The introduction of indexed NCDs would necessitate the maintenance of positive real rates for <u>all</u> loan rates of interest to prevent "round-tripping," i.e., borrowing at subsidised rates simply to acquire the attractive-yielding NCDs. The primary beneficiary of such tandem interest rate reform would be small and medium size business enterprises. Indirectly, of course, everyone would gain from the resulting lower rate of inflation and higher rate of economic growth.

The available empirical evidence mentioned earlier not only supports the case for a vigorous and flexible interest rate policy, but also indicates that economic behaviour in most countries is rather uniform. It does, therefore, allow some confidence to be placed in the application of international comparative studies to individual countries in the sample. At the very least, it makes the claim that any of these economies is different in certain important, fundamental respects more difficult to accept. Serious doubt is thrown on statements which pronounce that economic behaviour is insensitive, unresponsive or inelastic with respect to relative prices, be they between present and future consumption, domestic and foreign financial assets, imported and domestic consumer goods, domestic and imported factors of production, and even production techniques themselves.

The optimal interest rate policy would be the abolition of ceilings and the fostering of competition within the financial sector. The proposals outlined above call for large and dramatic changes in most of the sample countries, which may well be unacceptable in their entirety. Nevertheless, it is hope that the evident advantages of some of these measures will sway even the ardent supporters of selective credit policies and interest rate ceilings.

BIBLIOGRAPHY

S. Abe, M.J. Fry, B.K. Min, P. Vongvipanond and T.-P. Yu (1975), "The Demand for Money in Pakistan: Some Alternative Estimates," Pakistan Development Review, 14 (2), Summer 1975, pp.249-257.

S. Abe, M.J. Fry, B.K. Min, P. Vongvipanond and T.-P. Yu (1977), "Financial Liberalisation and Domestic Saving in Economic Development: An Empirical Test for Six Countries," Pakistan Development Review, 16 (3), Autumn 1977, pp.298-308.

B.B. Aghevli and M.S. Khan (1977), "Inflationary Finance and the Dynamics of Inflation: Indonesia, 1951-72," American Economic Review, 67 (3), June 1977, pp.390-403.

B.B. Aghevli and M.S. Khan (1978), "Government Deficits and the Inflationary Process in Developing Countries," International Monetary Fund Staff Papers, 25 (3), September 1978, pp.383-416.

Banco de Portugal (1963), Report of the Board of Directors for the Year 1962 (Lisbon: Banco de Portugal, 1963).

Bank Negara Malaysia (1979), Money and Banking in Malaysia (Kuala Lumpur: Bank Negara Malaysia, 1979).

Bank of Korea (1978), Financial System in Korea (Seoul. Bank of Korea, 1978).

E. Baltensperger (1972), "Economies of Scale, Firm Size, and Concentration in Banking," Journal of Money, Credit and Banking, 4 (3), August 1972, pp.467-488.

R.J. Barrett, M.R. Gray and J.M. Parkin (1975), "The Demand for Financial Assets by the Personal Sector of the UK Economy" in Modelling the Economy edited by G.A. Renton (London: Heinemann, 1975), pp.500-532.

K.B. Bhatia (1974), "Index-Linking of Financial Contracts: A Survey of the State-of-the-Arts" (Washington, D.C.: World Bank Staff Working Paper No.192, mimeo, November 1974).

V.V. Bhatt (1978), "Interest Rate, Transactions Costs and Financial Innovations" (Washington, D.C.: World Bank, Domestic Finance Study No.47, January 1978).

M.J. Boskin (1978), "Taxation, Saving, and the Rate of Interest," Journal of Political Economy, 86 (2, ii), April 1978, pp.S3-S27.

A. Brillembourg (1978), "The Role of Savings in Flow Demand for Money: Alternative Partial Adjustment Models," International Monetary Fund Staff Papers, 25 (2), June 1978, pp.278-292.

A. Brillembourg (1981), "Credit Policy in Indonesia" (Washington, D.C.: International Monetary Fund, mimeo, April 1981).

A.G. Chandavarkar (1971), "Some Aspects of Interest Rate Policies in Less Developed Economies: The Experience of Selected Asian Countries," International Monetary Fund Staff Papers, 18 (1), March 1971, pp.48-112.

H.-S. Cheng (1981), "From the Caymans," Federal Reserve Bank of San Francisco Weekly Letter, 13 February 1981, pp.1-3.

L.E. Christoffersen (1968), "Interest Rates and the Structure of a Commercial Banking System under Inflationary Conditions: A Case Study of Brazil" (Washington, D.C.: World Bank Economics Department Working Paper No.26, October 1968).

C.D. Datey (1978), "The Financial Cost of Agricultural Credit: A Case Study of Indian Experience" (Washington, D.C.: World Bank Staff Working Paper No.296, mimeo, October 1978).

D.S. Dutton (1971), "A Model of Self-Generating Inflation: The Argentine Case," Journal of Money, Credit and Banking, 3 (2), May 1971, pp.245-262.

Economist, "A Survey of International Banking," The Economist, 278 (7176), 14 March 1981.

M.J. Fry (1972), Finance and Development Planning in Turkey (Leiden: E.J. Brill, 1972).

M.J. Fry (1973), "Manipulating Demand for Money" in Essays in Modern Economics edited by J.M. Parkin (London: Longman, 1973), pp.371-385.

M.J. Fry (1974a), The Afghan Economy: Money, Finance and the Critical Constraints to Economic Development (Leiden: E.J. Brill, 1974).

M.J. Fry (1974b), Resource Mobilisation and Financial Development in Nepal (Kathmandu: Centre for Economic Development and Administration, 1974).

M.J. Fry (1976a), Portuguese Monetary Problems (Lisbon: Banco de Portugal, 1976).

M.J. Fry (1976b), "A Purchasing-Power-Parity Application to Demand for Money in Afghanistan," Journal of Political Economy, 84 (5), October 1976, pp.1133-1138.

M.J. Fry (1978a), "Alternative Stabilisation Strategies from a Model of Short Run Price and Output Fluctuations in Turkey," METU Studies in Development, 5 (18), Winter 1978, pp.21-36.

M.J. Fry (1978b), "Further Investigation into Turkey's Phillips Curve," METU Studies in Development, 5 (19), Spring 1978, pp.22-30.

M.J. Fry (1978c), "Deposits and Deposit Rates of Interest," Economia, 2 (2), May 1978, pp.261-283.

M.J. Fry (1978d), "Money and Capital or Financial Deepening in Economic Development?" Journal of Money, Credit and Banking, 10 (4), November 1978, pp.464-475.

M.J. Fry (1978e), "Pitfalls in Partial Adoption of the McKinnon-Shaw Development Strategy: The Nepalese Experience," Bangladesh Development Studies, 6 (3), Monsoon 1978, pp.257-270.

M.J. Fry (1978f), "The Money Supply Mechanism in Turkey," METU Studies in Development, 5 (21), Autumn 1978, pp.49-60.

M.J. Fry (1979a), "Monetary Control when Demand for Cash is Unpredictable: A Proposal for Stabilising the Money Multiplier in Portugal," Economic Journal, 89 (355), September 1979, pp.636-641.

M.J. Fry (1979b), "The Cost of Financial Repression in Turkey," Savings and Development, 3 (2), 1979, pp.127-135.

M.J. Fry (1980a), "Saving, Investment, Growth and the Cost of Financial Repression," World Development, 8 (4), April 1980, pp.317-327.

M.J. Fry (1980b), "Money, Interest, Inflation and Growth in Turkey," Journal of Monetary Economics, 6 (4), October 1980, pp.535-545.

M.J. Fry (1980c), "Mortgage Innovation?" Federal Reserve Bank of San Francisco Weekly Letter, 11 July 1980, pp.1-3.

M.J. Fry (1980d), "Money, Interest and Growth" in II Conferencia Internacional sobre Economia Portuguesa, 26 a 28 Setembro de 1979 edited by the German Marshall Fund of the United States and the Fundacao Calouste Gulbenkian (Lisbon: Fundacao Calouste Gulbenkian, 1980), pp.873-897.

M.J. Fry (1981a), "Interest Rate Policy in India" (Honolulu: University of Hawaii, Department of Economics, mimeo, February 1981).

M.J. Fry (1981b), "Interest Rates in Asia: An Examination of Interest Rate Policies in Burma, India, Indonesia, Korea, Malaysia, Nepal, Pakistan, the Philippines, Singapore, Sri Lanka, Taiwan and Thailand" (Honolulu: University of Hawaii, Department of Economics, mimeo, June 1981).

M.J. Fry (1981c), "Government Revenue from Monopoly Supply of Currency and Deposits," Journal of Monetary Economics, 7 (4), September 1981, forthcoming.

M.J. Fry (1981d), "Stabilization and Growth Strategies for Pacific Basin Developing Economies," Federal Reserve Bank of San Francisco Economic Review, Fall 1981, forthcoming.

M.J. Fry (1982), "Financial Development and Stabilisation Models for Financially Repressed Developing Economies," World Development, 10, 1982, forthcoming.

M.J. Fry and M.R. Farhi (1979), Money and Banking in Turkey (Istanbul: Bogazici University Press, 1979).

M.J. Fry and A. Mason (1981), "Children, Capital Inflows, Interest and Growth in the Life Cycle Saving Function" (Honolulu: University of Hawaii, Department of Economics, mimeo, May 1981).

H.G. Grubel (1977), "A Theory of Multinational Banking," Banca Nazionale del Lavoro Quarterly Review, (123), December 1977, pp.349-363.

Z. Hodjera (1978), "The Asian Currency Market: Singapore as a Regional Center," International Monetary Fund Staff Papers, 25 (2), June 1978, pp.221-253.

D.M. Jaffee and K.T. Rosen (1978), "Estimates of the Effectiveness of Stabilization Policies for the Mortgage and Housing Markets," Journal of Finance, 33 (3), June 1978, pp.933-946.

B.K. Kapur (1976), "Alternative Stabilization Policies for Less Developed Economies," Journal of Political Economy, 84 (4, i), August 1976, pp.777-795.

D.R. Khatkhate and D.P. Villanueva (1978), "Operation of Selective Credit Policies in Less Developed Countries: Certain Critical Issues," World Development, 6 (7/8), July/August 1978, pp.979-990.

D.R. Khatkhate and K.-W. Riechel (1980), "Multipurpose Banking: Its Nature, Scope, and Relevance for Less Developed Countries," International Monetary Fund Staff Papers, 27 (3), September 1980, pp.478-516.

Y.-P. Lee (1980), Inflation Hedges and Economic Growth in a Monetary Economy (Stanford: Stanford University, Department of Economics, Ph.D. thesis, 1980).

E. Lundberg (1964), "The Financial System of Portugal" (Washington, D.C.: International Bank for Reconstruction and Development and International Monetary Fund, October 1964).

I. McCarthy (1979), "Offshore Banking Centers: Benefits and Costs," Finance & Development, 16 (4), December 1979, pp.45-48.

R.I. McKinnon (1973), Money and Capital in Economic Development (Washington, D.C.: The Brookings Institution, 1973).

R.I. McKinnon (1980), "Financial Policies" in Policies for Industrial Progress in Developing Countries edited by J. Cody, H. Hughes and D. Wall (New York: Oxford University Press, 1980), pp.93-120.

A.H. Meltzer (1974), "Credit Availability and Economic Decisions: Some Evidence from the Mortgage and Housing Markets," Journal of Finance, 29 (3), June 1974, pp.763-777.

W.L. Ness (1972), "Some Effects of Inflation on Financing Investment in Argentina and Brazil" in Financial Development and Economic Growth: The Economic Consequences of Underdeveloped Capital Markets edited by A.W. Sametz (New York: New York University Press, 1972), pp.228-254.

S.J. Nickell (1977), "The Influence of Uncertainty on Investment," Economic Journal, 87 (345), March 1977, pp.47-70.

M. Olson and M.J. Bailey (1981), "Positive Time Preference," Journal of Political Economy, 89 (1), February 1981, pp.1-25.

E.S. Shaw (1973), Financial Deepening in Economic Development (New York: Oxford University Press, 1973).

E.S. Shaw (1975), "Inflation, Finance and Capital Markets," Federal Reserve Bank of San Francisco Economic Review, December 1975, pp.5-20.

C.G.F. Simkin (1978), "Hyperinflation and Nationalist China" in Stability and Inflation edited by A.R. Bergstrom, A.J.L. Catt, M.H. Peston and B.D.J. Silverstone (Chichester: John Wiley, 1978), pp.113-131.

V. Sundararajan and S. Thakur (1980), "Public Investment, Crowding Out, and Growth: A Dynamic Model Applied to India and Korea," International Monetary Fund Staff Papers, 27 (4), December 1980, pp.814-855.

V. Tanzi (1976), "Fiscal Policy, Keynesian Economics andthe Mobilization of Savings in Developing Countries," World Development, 4 (10 & 11), October-November 1976, pp.907-917.

V. Tanzi (1977), "Inflation, Lags in Collection, and the Real Value of Tax Revenue," International Monetary Fund Staff Papers, 24 (1), March 1977, pp.154-167.

F.A. Uluatam (1973), Monetary Multipliers and a Self Generating Inflation Model: The Turkish Case (Ankara: Turkish Republic State Planning Organization, 1973).

D. Usher (1977), "The Economics of Tax Incentives to Encourage Investment in Less Developed Countries," Journal of Development Economics, 4 (2), June 1977, pp.119-148.

G.J. Viksnins (1980), Financial Deepening in ASEAN Countries (Honolulu: Pacific Forum and distributed by the University Press of Hawaii, 1980).

Printed and bound by CPI Group (UK) Ltd, Croydon, CR0 4YY
09/11/2026
14999541-0001